Creativity and Technology

Creativity and Technology

Social Media, Mobiles and Museums

Edited by James E Katz,
Wayne LaBar & Ellen Lynch

MUSEUMSETC | EDINBURGH

Contents

Acknowledgements

This material is based upon work supported by the National Science Foundation under Grant No. 0610352. Any opinions, findings, and conclusions or recommendations expressed in this material are those of the author(s) and do not necessarily reflect the views of the National Science Foundation.

The international conference in the spring of 2010, out of which this collection of essays developed, was the product of a joint vision by Wayne LaBar, Ellen Lynch and James E Katz. We were inspired by a National Science Foundation initiative to consider the potential of mobile communication and social media in advancing informal science learning, especially as a way to reach underserved audiences and communities.

Denise Bressler and Kate Haley Goldman have been especially important in getting the project underway at its earliest stages. At the other end of the intellectual creation cycle, student note-takers from Rutgers captured in detail the conference proceedings; these in turn were fundamental to the production of the conference report that later laid the groundwork for this book. These students included Michael Barris, Steven Shie, Sun-Kyung "Sunny" Lee, Natise duBois, Mary Nucci, Michelle Riley, Rhaman Johnson, Daniel Halpern, Marian Stewart Titus, Yi-Fan Chen, Chih-Hui Lai, Yumi Jung, Hyunsook Youn and Ruthann Auten. Besides these individuals, other students who participated in the project were Frank Bridges, Amanda McGarry and Lora Appel. Frank Bridges not only assisted at the event itself, but was instrumental in helping create exciting graphics for the event. Amanda McGarry, with her tireless commitment to excellence, provided critical support with conference logistics and also was able to add her voice to the proceedings at an appropriate juncture. Lora Appel helped with conference arrangements and was instrumental in supervising the preparation of the chapters and manuscript; the editors are most grateful to her for her professionalism and dedication.

Others at Rutgers helped us connect the world of ideas to a sense of place. In this regard, we would like to recognize Suzanne Delehanty, director of the Zimmerli Art Museum at Rutgers, who not only generously shared her time and ideas but also helped make available the beautiful reception space at the Museum. Senior Associate Dean Hartmut Mokros, of the School of Communication and Information, provided much of the inspiration for the overall project and, during the conference itself, added provocative ideas about the importance of stability and tradition in physical space as well as reminded the attendees about the limits of technology in the museum setting.

Susanna Gibbons provided expert editing assistance, and Graeme Farnell deftly facilitated the publication process.

To all these people, we offer our heartfelt gratitude.

James E Katz
Wayne LaBar
Ellen Lynch

Foreword

EMLYN KOSTER

President and CEO
Liberty Science Center

The international discourse on progress and trends of mobile devices and social media in museums that took place April 26-27, 2010, in New Brunswick, New Jersey, USA, was the kind of focused, timely consideration that is unfeasible at multi-subject, association-level conferences of the museum field. I gratefully acknowledge its successful joint organization by Liberty Science Center and the School of Communication and Information at Rutgers University, its funding support by the US National Science Foundation, and particularly the background vision of Wayne LaBar and James Katz, PhD, and energetic planning by Amanda McGarry and Ellen Lynch of Liberty Science Center and the School of Communication and Information.

Especially in the sciences, the museum field over the last decade or so has undergone an accelerating evolution of its means and ends. *Means* and *ends* are instructively defined as efficiency and effectiveness, respectively "doing things right" and "doing the right things" as Peter Drucker, the management and leadership guru of the second half of the last century, clarified. Lately, the science museum field has been adopting new relevancy-driven approaches to be more effective (Koster, 2010) and the technological toolkit of the entire field has been rapidly diversifying with the aim of becoming more efficient (LaBar, 2010).

Museums, Mobile Devices and Social Media aimed to attract those at the vanguard of philosophy and practice in the integration of new communication technologies in museum exhibitions, programs and marketing. It remains important, though, to be mindful that museum users span the full range of ages and stages of learning and adaptation. Audiences therefore span a correspondingly wide range of ease with, and use of, the relentlessly advancing array of mobile devices and social media vehicles. Citing recent audience research in the United States, West (2010) offers a cautionary note – perhaps disagreeably so to a portion of this volume's contributors – that young adults may prefer that museums maintain traditional approaches for a respite from the dizzying pace of technological advance in the other dimensions of their personal and workplace lives.

Given the mounting volume of opportunities and challenges that face society and the environment – to which museums can, and arguably should, be usefully directed – the conference speakers and now the authors in this proceedings share a profound responsibility to strive toward efficiency innovations that do not compromise, but rather only ever strive to enrich, the greater potential of the museum's effectiveness.

A thoughtful synergy of the museum's efficiency and effectiveness has never been more vital.

Emlyn Koster, PhD
President and CEO
Liberty Science Center
www.lsc.org

References

Koster, E.H. (2010), Evolution of purpose in science museums and science centres. In: *Hot Topics, Public Culture, Museums*: F. Cameron and L. Kelly (Eds.), Cambridge Scholars Publishing, Ch. 4, p.76-94.

LaBar, W. (2010), A new museum paradigm for a changing world. *The Informal Learning Review*, No. 103, July-August 2010, p. 1-5.

West, R.W. (2010), Life stages of the museum visitor. *The Informal Learning Review*, No. 103, July-August 2010, p. 24.

Introduction

WAYNE LABAR

Liberty Science Center

Certainly, the amazing changes that have occurred in the museum world and the capabilities that the revolution in communication and information technology has provided our society could not have been foreseen just four years ago. Who knows where communication technology and design may be headed in the future?

This volume is actually a snapshot and record of what some of the leading thinkers, experimenters, and implementers in the museum, education, and media fields are currently wrestling with in terms of leveraging tools that allow us to communicate anywhere, at any time, and in new web-based social milieus. As you read through these papers, I believe you will discover a number of distinct yet prevalent themes that serve to interconnect their discussions and ideas.

The first is relevance. Mobile communication platforms and social media in a museum setting must be relevant to the public that use them as well as the institutions that are implementing them: whether it be relevance to a universal human subject, such as in an exhibition titled *Cooking*; or relevance to cultural heritage, as in Monica Murero's paper on embracing social media in museums; or the relevance of human rights as described by Corey Timpson in reference to a new museum project in Winnipeg.

The second is relationships. As might be expected, social media and mobile devices are platforms intended to encourage immediacy and communication. In essence, they are about creating relationships between people, and in our case, between people and museums and content. Papers such as those by Nancy Proctor, Theo Hug, and Richard Ling challenge the museum field to think differently about the relationships that they can create with their visitors and also the relationships that they can foster among individuals.

The third is response. At heart, mobile devices and social media are about being able to respond or engage interactively. With these new methods of engaging the public, new and impactful experiences are possible. New learning outcomes can be attainable as well. Anne Kahr-Højland, Martha Ladly, André Caron, and Letizia Caronia explore this theme in their project overviews.

I am certain readers will find other themes that connect all of the projects and experiences that are documented in the papers found in this volume. This is an exceptionally informative and exciting new arena for both the museum field and formal education. Certainly, our collective efforts are not without challenges; however, we share common goals of accessibility, engagement, and inclusion. As we learn more, explore more, and share findings, we continue to discover new ways in which these powerful tools can be used and enhanced to extend our missions and to deepen and broaden our impact on society.

I hope you enjoy this small taste of what was a lively and engaging two days of discussion, dialogue, and idea exchange. I personally wish to thank all of the participants in the conference, with a special thank you to James Katz and his students for their work in creating the conference and especially Ellen Lynch, Director of Exhibition Development and Operations at Liberty Science Center, who pulled our efforts together.

As one might communicate via mobile device or social media: TTYL!

Wayne LaBar
Vice President, Exhibitions and Featured Experiences
Liberty Science Center

From Headphones to Microphones: Mobile Social Media in the Museum as Distributed Network

NANCY PROCTOR

Smithsonian Institution

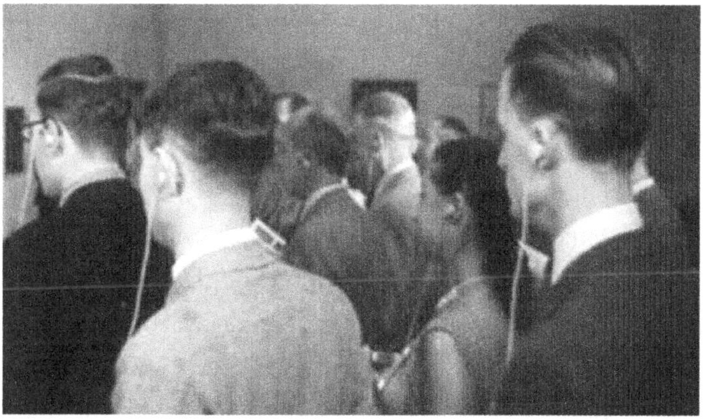

Figure 1: Still from a 1952 news clip from the Dutch Gescheidenis (History) website, demonstrating new audio tour technology at the State Museum in Amsterdam.

It's not about the technology.

In a black and white film (Fig. 1), a crowd of well-dressed Europeans moves as one through Amsterdam's State Museum. The voice-over explains, "By use of a 'hearing aid' the visitors get information and are being guided to the different artworks in the exhibition. The spoken words are recorded in several languages on a so-called tape recorder." As the *invisible guide* instructs, heads lean in to examine a detail in a painting, and then just as suddenly, move away to the right to follow the tour into the next gallery.[1]

Although this video from 1952, excavated by Loic Tallon, shows an example of one of the earliest tour technologies, a common perception of museum tours is that they are not terribly different today. Despite having evolved through numerous generations of mobile content delivery technology in the past 60-odd years, museum tours continue to face objections – however unfair they may be – that:

- Standardized tour content creates a homogenous, one-size-fits-all experience, possibly watered down to better appeal to mass audiences but ultimately boring both the novice and the expert;
- Museum tours dictate the visitor's movements and time in the galleries. Even though random access devices, which allow visitors to select commentaries in any order, have been prevalent in museums for nearly a generation, visitors still decline audio tours because they "prefer to visit at their own pace" (Luce, 2010). And it is true that wherever linear content is used – for example, a one-to-two minute audio commentary about an object – visitors can feel obliged to continue looking at that object until the end of the commentary, whether they remain interested or not.
- Digital tours produce herd behavior among visitors and crowding around exhibits as visitors look primarily, if not exclusively, at the objects featured on the tour;
- Audio and multimedia tours are in competition with human docents and represent an attempt to replace social interaction with a cold, isolating technology.

Perhaps partly as a result of this reputation, audio tours have never gained status as an indispensable part of the museum visit, despite having been introduced so long ago.

They are a *nice-to-have* rather than a *have-to-have*: after all, audiences do not visit the museum to take the audio tour. Indeed, take-up rates remain stubbornly well below 10% for most permanent collection tours, and many cultural professionals and members of the public are downright hostile to the platform (Longo, 2010). By contrast, in the past decade, one new technology after another has been heralded as the must-have tool for the next generation of museum interpretation: multimedia tours, cellphone tours, podcasts and downloadable audio tours, even Twitter and text-message tours – all have come on the scene amid high expectations, but have so far failed significantly to transform the interpretive landscape. iPhone and web apps are the latest great hope, and offer exciting new ways of reaching audiences on-site and beyond. Yet according to a 2006 study at SFMOMA, a minority of visitors uses technology for on-site interpretation, opting instead for traditional analog tools – if any – when given a choice (Samis, 2007). So are these new platforms doomed simply to replace the traditional audio tour with increasingly complex and expensive, but no less marginal, solutions for museum interpretation?

It's in the context of this repeating cycle of dashed new technology hopes that I'd like to propose we frame the question differently. Mobile interpretation is not about the technology: which player, which platform, which app, or other bells and whistles. It may be easy to assume that the Dutch museumgoers move as one in the 1952 film clip because the audio tour uses a broadcast technology: everyone is hearing the same content at the same time, so they are all compelled by the technology do the same thing at the same time. But imagine, for a moment, that the content asked visitors to spend a couple of minutes looking around the gallery, then to choose a favorite artwork and describe it to a companion. Using the same, archaic broadcast tour technology, we would have seen a very different experience played out in the museum. Instead of herd movement, we would have seen individuals experiencing the exhibits at their own pace. Instead of tomblike silence, we could have seen lively social interactions. Instead of obedient, passive visitors, we would have seen proactive teachers and ambassadors, advocating for the works that most touched them – reinforcing their memories, learning, and experiences by sharing them with others. When the content and experience follow the technology, instead of the other way around, we risk turning our visitors into comical cyborgs – a melding of human and machine – that follow borg-like behaviors prescribed by the platforms and interfaces we offer them. If the museum tour experience has not made many gains in popular opinion since 1952, despite all the innovations in mobile technology in the past 60 years, perhaps it's the content, not the technology, which is the problem.

The audience takes the lead

A compelling interpretation program is the result of great content and user-centric experience design first and foremost. And in the *www* age of *whatever, whenever, wherever* information on demand, our visitors are increasingly determining the museum's interpretive experience themselves, using new informational practices adapted from their Web 2.0 lives. Since we can no longer even assume that our visitors are physically in the museum, the question becomes: how does the museum "meet them where they are – and take them someplace new" as Michael Edson, director of Web and New Media Strategy at the Smithsonian Institution, has asked?

Letting our audiences lead our mobile program design means thinking about content and experience design first and foremost. Following are six questions towards the design of an evaluation-led, user-centric mobile experience. But before embarking on any mobile project, it is worth any interpretation designer's time to check out the research and evaluations of previous mobile projects and their audiences. By 2005, over 100 handheld pilots had been conducted at museums around the world (Proctor, 2005). Most of these repeated some, if not all, prior experiments in ignorance of their colleagues' learning in other museums. Although convincing internal stakeholders that evaluations are worthwhile can be a challenge, significant progress has recently been made in centralizing access to what data has been published on mobile platforms and projects. Several wikis and websites are referenced in this paper and make an excellent starting point for any mobile interpretation project.

At the same time, it is important to take into account the radical transformations that recent technology and social media in particular, have wrought on contemporary culture. Concepts and applications that may not have worked even just a few years ago may now have currency in today's more mature Web 2.0 zeitgeist. And audiences are not always reliable predictors of what they want, like, or will use; sometimes we just have to try out new ideas in order to understand if they will work with our target audiences. For this reason, an appreciation for the value of audience-led design and basic understanding of the principles of prototyping and usability testing are critical skills for museum professionals in this era of rapid technology change.

1. Who is your target audience – really?

Whether you're targeting on-site visitors or virtual visitors through the mobile web, identifying and understanding your target audience is not just a question of demographics. In designing an interpretation program, we also need to know what platforms they already use to get content and information on a regular basis:

- Are they already in the habit of taking traditional audio tours?
- Are they comfortable using cellphones or smart phones?
- Do they regularly download podcasts and other content for time-shifted playback on their personal media players?
- Do they use mobile social media applications, such as text messaging (SMS), Twitter, Facebook...

If museums' visitors are reluctant to use technology in the galleries, web and cellphone use statistics show that they are certainly using it everywhere else, with internet penetration rates well over 50% for Europe and 74% for North America[2], and mobile phone subscriptions surpassing 60% globally.[3] What do they like about the tools and devices they use every day? What makes accessing content on those platforms compelling, addictive, and automatic – or difficult and not worth the effort?

In the light of your organization's mission and key messages, what do you want this target audience to know, think and/or feel? Having understood a bit more how your target audiences operate – what delights and frustrates them – we begin to have a sense of how to reach them on all the platforms they use, both inside the museum and beyond. And with a bit of forethought and judicious content design, it should be possible to create a mobile program that will move cross-platform as easily as your visitors do.

2. What do they want to know?

Just as important as understanding what mobile platforms and tools visitors already use and enjoy, is getting a sense of what they want to know. What in your exhibition or collection makes them curious? What questions pop immediately into their minds when they enter the gallery or encounter a particular object? Do existing interpretation solutions already satisfy those questions?

At the Smithsonian American Art Museum and the Indianapolis Museum of Art, museum studies students and interns experimented with "question-mapping" in the galleries. On maps of the museums, they wrote down the questions that came to mind as they toured the galleries, noting the point in the museum where the question occurred to them. Their questions ranged from the general to the specific: from questions about why the collection existed, to the curators' intentions in a given gallery hang, to questions about specific artworks' histories and meanings. This semi-structured interview method yielded rich insight into what inspired these visitors to stop, notice, and think about what they were seeing. It also revealed where existing interpretation was insufficient to satisfy and further stimulate visitors' curiosity.

Question-mapping is not the only way to gather the questions that structure the mobile interpretation program. Most museums get comments from visitors, written on cards, in books, or by e-mail, and posed in person to staff. These questions and "FAQs" are a rich resource for understanding what engages visitors in the collections.

Of course, it's easier to gather questions about an existing installation or body of content that is already available to the public online; but this isn't to say that mobile interpretation planning shouldn't start until the installation or web publication is public. On the contrary, the earlier the interpretive program is woven into the presentation design, the more effective it will be in terms of both impact on visitors and cost. When planning new displays and their interpretation, the evaluation with target audiences may take on a different form to determine their interpretive interests and needs for the proposed presentation, but it is no less critical to the overall planning process.

By starting with the visitor's questions rather than the curator's key messages, we enter into a conversation with our audiences, rather than a lecture. Even if aimed at a well-defined target audience, quite often the design of a mobile interpretation program starts by asking, *What are the key messages of this exhibition/collection?* Curators or other subject experts are consulted about what they really want visitors to leave knowing, and the tour is developed around this agenda. Little surprise, then, that audio tours have so often missed the mark when their intended audiences aren't even consulted on what content they're interested in having.

3. Where do you want to take them?

Leading the mobile program design process with the visitors' questions makes us think about the needs of different kinds of visitors: what do first-timers and novices to the field of study represented by the collection or exhibition want to know, versus what interests a more experienced visitor? Sometimes visitors may not be familiar enough with a subject even to think to ask certain questions; or the exhibit is "like Teflon" and requires "interpretive Velcro" to hook the passing visitor (Samis Velcro, 2007). Here the input of curators and other experts is key: by engaging and rewarding the visitors' most general and fleeting of initial interests, an opportunity is created to capture their attention and introduce them to other aspects they may not have thought to question. Now that we have the visitors on board, where do we want to take them? The subject expert is in the best position to chart this course.

In finalizing the map of the visitors' questions and the new insight that the curatorial team will add to that conversation, it is probably easiest to follow the organization of the collections or exhibitions: both physically, for on-site visitors, and conceptually or by object, for those who may be visiting through the museum's website or other digital platform. But we should also think outside of those traditional boxes, based as they are on curatorial ontologies that may or may not be compelling, or even visible, to non-expert visitors. Can we use visitors' questions to conceive of other ways of organizing information in our collections?

Once grouped into whatever broad categories are used for structuring the mobile content and experience, it will generally be necessary to filter the responses by questions that speak to the museum's mission and key messages for the target audience – so that resources are put towards answering the questions that will yield most interpretive benefit. Above all, we should prioritize the questions and topics that elicit great stories, as they are what visitors will remember most of all about their museum visit.

4. Soundtracks, sound bites, both or more?

In the early days of audio tours, storytelling was actually

facilitated by the limitations of the technical platform. Whether given by live guides, broadcast as at Amsterdam's State Museum in 1952, or lugged through the museum on reel-to-reel tape players, the first museum tours were linear:

From starting point A to end point N, the exhibits interpreted on the tour were strung along the tour's linear route like pearls of wisdom on a necklace:

-o-o-o-o-o-

The value of the tour was measured in stops:

o

The messiness, but also the magic, happened in the spaces between the exhibit commentary or "stops" on the tour:

-

People got lost in the interstitial spaces, uncertain of where to find the next stop (o) – or lost track of where they were in the audio tour tape:

-o-o-o-?

Or they got bored, distracted, or tired of following the herd – or simply decided to get off the tour at the last stop:

-o-o-o~§

But in the best linear tours, the spaces in-between were where it all happened – that was where you got the background information and context that brought the exhibits to life:

-o!o!o!

The connective tissue of the tour immersed you in music and storytelling that carried you along effortlessly from one stop to another, transporting you to a different, magical world. In some courageous tours, the liminal spaces were an opportunity for audience participation. The tour could issue challenges to the visitors to play games, complete tasks, or simply give time to share impressions with a companion. The more marketing-minded tours gestured toward galleries and exhibits along the way as opportunities to return and find out more in another tour:

-o+o+o+

Fear and impatience with the messiness prevailed, however, and the second generation of audio tour technology introduced "random access" tours. Visitors could choose which exhibits they saw and hence which "stops" they listened to absolutely at random:

o o o

o oo o o

o o

It was the first digital *personalization* in museum tours; ambitious vendors promoted it in Modernist terms as a liberation from the herd, eager to recoup their investment in the new technology and steal a march on the competition. Thanks to the new technology, we could finally do our own thing in the museum.

But as Laura Mann, a veteran of the audio tour industry, once remarked, "We lost something when we moved away from the linear tour." (Mann, 2009)

By focusing on content and experience design rather than simply bringing out a better audio tour technology, I think we can recover some of that early magic – and even some of the productive messiness – in the next generation of museum tours.

Soundtracks: A variant on the original linear audio tour, the soundtrack provides tools for understanding key principles of the displays. It helps me avoid being blinded to the forest by the trees. When I walk into an exhibition space, the first thing I need to know may not be when that painter was born and died; I probably need some conceptual orientation: where am I? What's going on here? Why is all this stuff in this one place? What am I supposed to be taking away from it? The typical two-minute *Director's Welcome to the Exhibition* message, in which the marketing blurb for the show is dryly reiterated, is unlikely to do the trick. The visitors have already read that. They've bought the ticket or committed time to entering the gallery. Now we need to give them the keys to unlock the magic they find there. Help them see what the curator sees: the stories that connect and hold the displays in an electrifying web. Show them the sparks that fly when you put those artists in the same room, or on the same web page. Directly or indirectly, explain the process that produced the connections among selected exhibits, the color of the walls, the type fonts. Call their attention to how the installation inflects, seduces, and stimulates their vision, their curiosity, and their understanding. Give them the tools to rediscover the display's main concepts wherever they may go after the museum visit. After leaving a Picasso exhibition, I should have some grasp not just of the Picassos that were in that show, but of any Picasso I may encounter hereafter. I should understand why the exhibition's scientific or historical concept is important, and read it into my everyday life. The aim here is not just some pedantic, pedagogical mission: recognition and understanding produce pleasure in the visitor, which is why we so often seek out exhibits and artworks we know and pass by the unfamiliar to revisit old friends.

For those visitors in a hurry, or uncommitted to more than skimming the surface of this exhibition at first, the soundtrack overview is ideal: packing meaning into a survey of a number of objects, thereby increasing the chances of the visitor getting hooked on one and staying longer to learn more. The soundtrack can also be designed as a downloadable or fixed web tour, meaningful for audiences off-site as well as in the gallery. Especially if illustrated as a video or slideshow, the overview soundtrack provides, through the web, access for all those non-visitors who'll never be able to come to the museum in person, and sometimes far outnumber the actual on-site visitors at some museums (Griffiths, 2006). Our websites are designed, presumably, to serve those who'll never be able to visit us physically, as well as those who are planning a visit or researching an interest sparked on-site; why shouldn't our mobile content also aim to achieve that broad reach and return on investment – especially since now an increasing amount of our traditionally fixed web activity, like e-mail and web surfing, is done on a mobile device? Indeed, Gartner predicts mobile devices will overtake PCs for Internet access by 2013 (Leggatt, 2010).

Ideally, the soundtrack is a story or a conversation that

the visitor can join. It can be a rich narrative in which the visitors immerse themselves, or a dialogue that opens up a space for the visitor to join the conversation and form his or her own opinions. As demonstrated by Beth Harris and Steven Zucker through SmartHistory.org, the dialogue models how the experts acquire their knowledge, and it can be more engaging than the straightforward lecture in which the visitor is positioned as the passive recipient of wisdom from on high.[5]

Although much longer than the more familiar audio tour stops, the soundtrack can be divided into a number of segments without losing the plot, because the segments are connected and read sequentially. One logical way of creating chapters or shorter segments from the soundtrack, to facilitate download and podcasting, is to divide the soundtrack by room or gallery overviews. Alternatively the story may be divided thematically, as in an account of the story of American art by medium, period, or movement. One benefit of the audiovisual interfaces now available on new handheld platforms is that there are several ways for visitors to relocate themselves, both in the galleries and in a piece of linear content, should they deviate from the linear path at any point – from interactive maps to illustrated audio stops that serve as visual orientation aids.

Sound bites: Sound bites are the curated stops, or nuggets of information, on specific objects or exhibits. Their most common interpretive analog is the wall label. Since the introduction of random access digital tour technology, the majority of audio tour experiences have been based on the visitor choosing which exhibits they want to hear about, and then triggering the commentary by dialing in the related stop number on their audio play. Often those commentaries are audio versions of the object label, rewritten relatively well for the ear. The fact that visitors often lose patience with standing in one place for the duration of the stop has prompted Beth Harris to refer to these commentaries as *starts*: visitors start hearing the message while looking at the related object, but walk away before the end (Harris, 2009). Alternatively, stops are too short, failing to go into enough depth about the object that really captivates you, nor answering your burning question about it.

There is no such thing as one-size-fits-all interpretation; like clothing in a variable climate, stops should be layered, and ideally enable access to a variety of types of content to satisfy different kinds of questions and learning styles: not just audio and video clips, but wall labels, catalog text, interactives, and user-generated content too. Uniformity may be neat and comforting but it's also boring. By varying what's available, we set an achievable goal in content production for the museum: if you've got audio, video or multimedia commentary, great; if not, make your collection records, wall text, and catalog essays available; and link to others' content, visitor commentaries, favorites, and votes. Tate Modern tested visitors' reactions to mixed media and style content on its *Collections Tour* of 2005 and found that while they certainly preferred richer media, they didn't mind simple text and collection records if that was all that was available; what they hated was finding nothing at all (Wilson, 2004). With more varied and layered stop content, Tate also introduced an element of mystery and surprise to each object's stop menu.

The ideal interface for soundtracks, sound bites, and more: In theory, sound bites are like atoms that can be searched and combined with other stops in a variety of ways to build new insights and bodies of information around a given topic. In practice, I have not yet seen this achieved on a large scale in digital tour programs, where it seems sound bites are almost impossible to get right as stand-alone units of information. This is why severing them from the linear narrative of the early audio tour had such a deleterious effect on the ability of the random access tour to satisfy visitors' curiosity and innate desire for narrative.

With the right interface, sound bites can be reattached to the soundtrack at relevant points along its timeline, and equally can link to third party and user-generated content to serve as a conduit to even deeper and broader content. By the same token, when used in strategic combination with links to sound bites and interactive functionality, the soundtrack can both trigger and capture visitor response – e.g., through voting, favoriting, or leaving comments as text (including tweets and e-mails) or voice (via cellphone).

The video interface developed by ArtBabble.org probably provides the best model for successfully combining soundtracks, sound bites, and links to other content and applications. In this design, the soundtrack overviews can be presented as videos in which the sound bites and links are

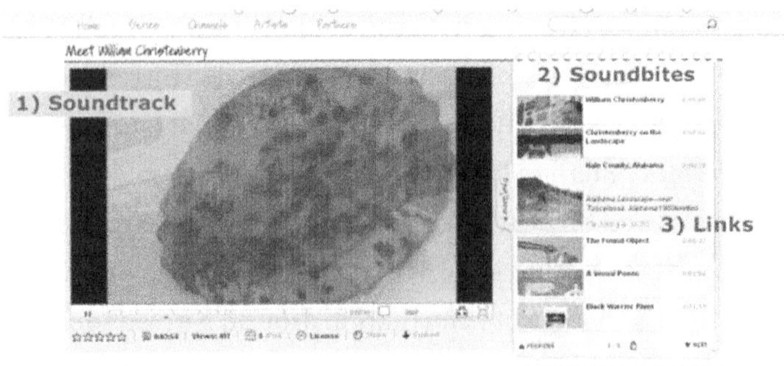

Figure 2: An ArtBabble interface, annotated to indicate the sound bite, soundtrack and linked content. By Shelley Mannion.

embedded as notes. Visitors can choose whether to follow the linear narrative of the tour (either on-site or online); at whatever point, object, or display that their interest is piqued, they can drill down to a stop or link to related third-party content.

Soundtracks, sound bites, and other content like interactives can be locally stored on a tour device for use in an offline environment (i.e., without Wi-Fi or cellular network coverage); links to content and applications in the cloud, including social media and feedback mechanisms, require a network connection. The physical and technical possibilities of the museum can therefore determine the extent to which third party (non-museum) and user-generated content and applications can be incorporated into the on-site mobile experience. On the other hand, for virtual or non-visitors, this kind of social media practice is by now a given. As we design mobile experiences that are meaningful beyond – as well as within – the museum's walls, these considerations of how

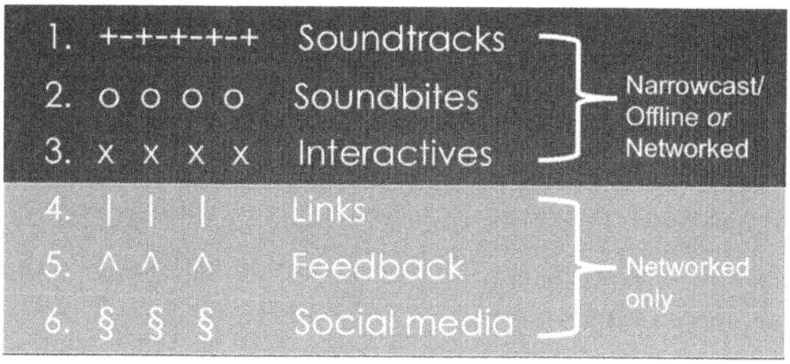

Figure 3: Some mobile content can be locally stored for playback in an "offline" environment. Content and applications hosted "in the cloud" require cellular or Wi-Fi connectivity.

our audiences consume content and information in the rest of their lives put increasing pressure on the museum to ensure ubiquitous network connectivity in order not to cut visitors off from their usual informational habits and pleasures.

5. How will you tell the story?

There is little media more intimate and personal than mobile. As Jonathan Finkelstein has remarked, speaking into someone's ear, or interacting with him on his personal device, is about as close as a museum can get to its visitors (Finkelstein, 2009). The power of that proximity should be matched with careful selection of the voice trusted to convey the museum's message: someone who is at once insightful about the subject, able to relate the content to the museum's mission and key messages, and is a good communicator with the target audience in terms of voice and manner.

By far the most requested tour narrators are the artists, curators, or other experts who are perceived as authoritative on the subject at hand. After all, visitors come to the museum and take guided tours to get accurate and useful insight into the topics addressed by the collections and exhibitions. This optimistic engagement on the part of the visitors is quickly squandered if the expert voice or the content is hard for the visitor to connect with: not relevant or accessible to the audience's interests.

Equally common in traditional museum tours are professional narrators. Although they often bring higher production values to museum audio content than non-professional speakers, the anonymity of professional narrators can jar with audiences increasingly used to the personal touch of our social media age. For many audiences, it's easier to connect with an interlocutor who is named and whose relevance to the subject matter and the audience is clear.

Interviews of curators and experts by a friendly interlocutor have been a frequent technique in audio tours for establishing a less formal and more accessible tone, while cutting production costs by removing the need for scripts. SmartHistory.org has shown us how to open the interview up further into a dialogue in which both parties are exploring and learning about the exhibits and objects on view together. This structure seems to create a space for the listener to participate as a third point of view, and models the processes that create subject expertise.

Some museums, including, notably, Tate Modern and SFMOMA, have also included *vox pops* – interviews with visitors or other non-experts – in their mobile programs to add the interest and relevancy of multiple voices to the museum's own more authoritative or analytical content. These early experiments with user-generated content have indicated the potential to empower the visitor as a storyteller. By equipping visitors with the means to capture their own impressions and learning from the museum to share with others, we turn them into a global teaching force that will act as advocates for the museum for generations to come.

6. What platform(s)?

Once we have addressed all other aspects of the content and experience design, we should focus on the technology so that it serves the needs of the visitor and the experience, not *vice versa*. This table summarizes the strengths of each platform in supporting each kind of content: playback content like soundtracks and sound bites; content using single player interactive interfaces; and networked content and applications, including links to third-party content, social media applications, and user-generated content and feedback.

Museum-purposed audio and multimedia players: Even in a ubiquitously connected world, there will always be visitors who come to the museum without a mobile device to use, or with a flat or short-lived battery; they therefore will need to pick up something on site to enjoy the tour and available mobile services. This number, and hence staff and hardware overhead, can be minimized by publishing to as many other

	Audio player	Multimedia player	Cellphone	Personal media player	Smart phones	
					Mobile Browser	Mobile App
Soundtrack	x	x	(x)	X	X	X
Soundbite	X	X	X	x	X	X
Interactive		X			X	X
Link			X		X	x
Feedback			X		X	X
Social media					X	X

Figure 5: Platform support for mobile content types.

platforms as possible, and by marketing downloads, through pre-visit channels: for example, the visit information on the museum website should loudly encourage visitors to download the podcast tour or iPhone app prior to visiting the museum. As discussed in the *MuseumMobile* podcast interview with the principals of SmartHistory.org, visitors prepare for other trips – plane journeys, the daily commute, going for a run – and we need to teach them to think of the museum in the same way.[6]

Podcasts and other downloadable content: Podcasting and its video equivalent, vodcasting, should be a standard additional means of distributing tour content – and not just audio and video but also related PDFs. Both soundtracks and sound bites can be easily provided either freely or for a fee through the iTunes store and iTunes U (where all content is free), as well as from the museum's own website and other RSS outlets. Downloaded content can be used in the museum or anywhere the visitor chooses to take it.

Though it's difficult to measure what subscribers have listened to, podcasts offer a powerful and personal channel through which to develop a relationship with museum audiences and push new content on a regular basis. In addition, because the content is stored locally on the visitor's own media player, it can be enjoyed even if no network connectivity is available – in the museum, on the visitor's daily commute, or wherever he or she may be inspired to enjoy the production.

Once subscribers have signed up for a podcast or RSS feed, it's easy to offer new information of varying types through the feed as well, including invitations to enjoy further museum programs, or to support the museum with a donation or membership. Many NPR (National Public Radio) programs have begun with a call for donations in their podcasts, using mobile giving through text message to make sponsorship easier and better capture listeners' generous impulses in the moment.

The cellphone: Cellphone use is not always possible in museums, either because of poor reception or cultural opposition to using phones in the galleries. Even where cellphone tours can be used, the audio quality is poor, and for many, holding a phone to the ear for a traditional 45-minute museum audio tour would be exhausting – though cellphones with hands-free headphones are clearly much more comfortable in this respect. Other visitors are loath to run down their cellphone battery taking an audio tour if they're going to be out all day (Whitney, 2009), or simply prefer other tour platforms. There are also business model limitations to the cellphone tour: it will really only ever reach a domestic audience, and with per-minute fees charged to the museum, the cellphone tour can actually be a victim of its own success if more people call in to the tour than budgeted for.

But cellphones are a great solution for *audio snacking* – quick calls to find out about a handful of objects that pique our interest, especially where the museum is not able to offer mobile devices on site. Perhaps most powerfully, cellphones can be used as two-way communication devices (Proctor, 2009a). I can listen to a stop, take a quiz, vote, and leave an audio or text message comment through my phone. Plus the phone is mine, so the museum doesn't have to train me on its use and interface. By thinking of the tour experience as more participatory or game-like, museums can create great mobile experiences for domestic audiences' cellphones that will take advantage of the best of both the traditional audio tour and the two-way communication device, using the platform that has reached the largest percentage of the world's population.

Mobile websites: By 2009, 32% of Americans had accessed the web through their mobile devices.[4] As we use our mobile devices to do more of the surfing previously reserved for desktop or laptop computers, museums' mobile offerings become increasingly visible and in greater demand. A web-based version of museum tours, designed for the small screens of smart phones, not only increases access to the content for on-site visitors, but also for non-visitors – they can virtually tour collections and exhibitions, thanks to soundtrack and other content that can be meaningful independent of the physical site.

Mobile websites require Wi-Fi internet access or a high speed data network (for domestic audiences), but where these are available to visitors on- and off-site, they can be a cost-effective platform for the museum and visitor alike. More easily designed for use by a wide range of smart phones than dedicated apps like iPhone or Android apps, mobile web-based content and applications are free to end users wherever free Wi-Fi is available, regardless of whether their network plan is with a domestic or foreign network provider.

Mobile web, like smart phone apps, also provides the opportunity for the tour to incorporate or reference other functions available on the mobile device, such as leaving voice or text comments, taking pictures, or sending content from the visit or tour to friends. Koven Smith has written an insightful paper about expanding our thinking on the mobile museum experience beyond the "tour" paradigm to incorporate the major mobile web functions and applications that people already use daily (Smith, 2009).

Native or Smart phone apps: The platforms chart (Fig. 5 above) makes it clear why developers and museums are so excited about smart phone apps. Combining the versatility of a web app with the stability of locally-stored content, smart phone apps support the full range of content types that can be included in mobile interpretation programs today.

But currently smart phone apps require bespoke

development for each mobile phone platform and are limited to the small, albeit growing, number of people who own smart phones. And although Wi-Fi hotspots in the museum can permit downloading of apps – even apps that incorporate data transfer on-site (e.g., streaming some content during the visit, or transmitting messages, votes, etc. from the tour) – not all museums are able to provide strong or ubiquitous Wi-Fi, so app use may be limited to those who have the time and foresight to plan their visit and download the app before visiting the museum. Nor should it be assumed that visitors will have enough free space on their smart phones to download an app, especially if it's heavy with content, without deleting other content from their phones first. Nonetheless, there is no question that dedicated smart phone apps are a growing platform for museum content, both for on-site tours and for extending the museum visit beyond its walls.

Thinking outside the audio tour box: As this wide variety of mobile platforms demonstrates, today not just audio tours but mobile multimedia, including text, video, and interactive content, can be delivered to the visitor's own devices and to devices provided on-site by the museum. Museums can "meet our audiences where they are" by publishing museum content in as many places as possible, taking their collections, exhibitions, and expertise to the digital communities and platforms their target audiences already frequent.

The multi-platform museum: In a multi-platform museum practice, traditional audio tour and even web content are repurposed and published *cross-platform* to a range of devices, ideally with some planning ahead and versioning for the different delivery platforms: cellphones, personal media players like iPods, and web-enabled phones, as both mobile websites and native applications. It sounds like so much low-hanging fruit.

At first glance, *multi-platform museum* is a perfectly serviceable term to name this multi-faceted museum interface: it makes us think beyond the physical site and suggests the museum's presence in multiple digital contexts. But for both the content designer and the technologist, *multi-platform* has implications that make another metaphor, the distributed network, a more engaged way to approach 21st century museum outreach across digital and other platforms.

Technically speaking, *multi-platform* implies publishing to many outlets, or platforms, from a single content source. In other words, the aim is to create a direct copy – as perfect a replica as possible – of the same original (content, message) on each platform, or at least control the content and experience centrally. For the content developer, this means economy of scale: build once, publish many times. But like any wholesale export of culture without sensitivity to the "native" context and its communities, multi-platform publishing results at best in forcing square pegs into round holes, at worst in a sort of colonizing effort; either way, it usually fails. Content designed for one use, context, or platform rarely ports directly and easily onto another. Brochures do not make good websites. Texts written for catalogs and wall labels sound stilted and dry as audio guide scripts. And just as physical artifacts have to be photographed or scanned in order to create a digital representation for online use, at a minimum videos and photographs have to be recoded and reformatted to meet the size requirements of social media platforms.

And as Koven Smith has asked, will broadcasting what is fundamentally the same old content to shiny new gadgets really take our audiences "some place new"? (Smith 2009) Will adding images, video, or a sexy new consumer device really improve the take-up or penetration rates of audio tour use by museum visitors? Although in conflict with visitors' self-reported use of mobile interpretation in museums (Petrie 2010), in reality the audio tour reaches a sobering minority of the museum's on-site audience; one has to ask how well podcasts, cellphone tours, and iPhone apps serving up the same basic content fulfill the needs of those audiences who'll never visit the museum in person. And what museum has the resources to develop content and experiences that will suit the full range of its audiences on all the mobile platforms?

Tailoring content and experiences for each audience and platform – seemingly the alternative to a mass-export approach to content publishing and management – can be prohibitively expensive, especially for reaching niche audiences. As a result, museums often live within their budgets by trying to develop a single, one-size-fits-all solution that is intended to serve a "broad" audience but in effect compromises both quality and specificity at the risk of fitting none. And furthermore, because of its spoke and wheel structure – pushing content out from a central point – the multi-platform model does not easily incorporate feedback loops, or ways for user-authored content and experiences to get combined with museum content and redistributed, without passing through centralized channels where editors and censors monitor the conversation.

Creating content and messages with a more discursive dynamic and greater sensitivity to the specificity of platforms, audiences, and contexts requires a different approach than what is implied by the multi-platform metaphor. Instead, we need more flexible, modular structures and methodologies for content and experience development, akin to what a technologist might call a *distributed network* or rhizomic model.

The rhizomic museum: There is no center or panopticon to a distributed network or rhizome, no Hegelian master or slave. The constituent platforms work together to create a whole that is greater than the sum of its parts. In terms of content, we speak less of the *original* and more of the *simulacrum* – the copy without an original, about which Jean Baudrillard wrote so eloquently (Baudrillard 1988). Notions of authority and hierarchy are not very helpful in describing relationships and processes that work together more like mash-ups than pronouncements. Truth, rather than being disseminated outwards from a center point, is discovered in its intersections and interstices, through the (sometimes surprising) juxtapositions that can happen when experiences are assembled collaboratively along the many-branched paths of a rhizome. In the Museum as Distributed Network, content and experience creation resembles atoms coming together and reforming on new platforms to create new molecules, or choose-your-own-ending adventure stories.

The internet is a distributed network, structured as a rhizome by its military inventors precisely to make it impossible to destroy the entire network through an attack on any single center – just as it is impossible to eliminate crab grass from one's lawn by pulling up a single plant! In the loosest sense, every time you perform a Google search, you have tapped into a distributed network. The original content that forms the results page – itself a mash-up – exists on a number of different computers, physical and virtual, each potentially in its own network or grid. It is delivered to you in an entirely new and personalized context on the basis of your search interests and terms at that precise moment in time; yet the original digital assets are completely untouched in their original form, and you can easily trace the results content back to discover its full context and original publication environment. The value and authority of the *original* are not diminished, but rather enhanced, by being placed in new contexts alongside content from other sources. Quality is determined by the content's relevance, depth, longevity, quantity of peer-reviews, and recommendations – also known as links.

Other than the internet itself, distributed networks don't really exist yet in the technical sense, because most computer systems are built on the master-slave model. But we do see the products of master-slave systems being distributed, for example in peer-to-peer gaming, social media environments, and some wikis. As we begin to design interpretation and information systems for the museum in the age of social media, the distributed network can serve as an inspirational metaphor to describe new ways of authoring and supporting museum experiences that are:

- conversational rather than unilateral
- engaging rather than simply didactic
- generative of content and open-ended rather than finite and closed.

They become "smarter", more effective and useful the more they are used: rather like Pandora or the Amazon recommendations system.

Like the internet, the Museum as Distributed Network is enhanced, not diluted, by multiple voices and authors. But it requires powerful tools for making the ever-increasing data and metadata of assets, interpretations, and interpreters findable, and for connecting communities of interest in meaningful ways across oceans of content and contributors. Here is one vision of how a distributed network approach can be deployed to foster and support mobile experiences of the museum both on-site and beyond.

Principal ingredients of an un-tour: As was discussed earlier, thinking outside the audio tour box – designing an *un-tour* – begins with asking what audiences want to know about our collections, exhibitions, research, and scholarship (both on-site and beyond the museum's walls) and not with what we want to tell them. By responding first to the questions that visitors ask of the museum, we are able to use mobile solutions to engage them more fully and immediately, and with longer-lasting effects. Most importantly, we can help connect communities of interest and facilitate conversations among our audiences as well as with the museum itself, turning the mobile experience into mobile social media.

1. *Data capture and feedback tools*: The foundational functions of the un-tour therefore include tools that capture data and feedback from visitors on where they go, what they see, and what questions they ask. Analytics and metrics reporting on their usage are key to understanding where further support, in the form of content or other responses from the museum, are required to reply to and further stimulate visitors' interest. Search facilities for collections and exhibition content, as well as interactive maps, can be important interfaces to help visitors find the content that responds to their questions and interests. Most importantly, the un-tour should include commenting, collecting, and sharing functionality that enables visitors to save, annotate, and redistribute their experience and insights to others. For those who may not be prepared to register their comments on a museum's displays for all the world to see, Facebook-style functionality that allows visitors to select the audiences who can see their contributions to the mobile conversation, can be an important feature to help them join the conversation in a comfortable and "safe" way.

2. *Facilitating conversations within communities of interest*:

The museum can help communities of interest form around objects and exhibits by facilitating the conversation, both directly with individuals and among audience members. The subject expert, who speaks through the museum to respond to questions or correct misinformation, is both just one voice and a privileged node in the peer-to-peer network of the conversation. The museum's subject experts are uniquely placed to identify and leverage opportunities in the community's conversation to take the participants to new levels of understanding, and to introduce new concepts and content that otherwise they may have remained ignorant of – without shutting down conversation by playing the role of *sage on a stage*. In this respect, the value and intelligence of the conversational network increases with use. It's therefore important for the museum not to be stingy, or to skimp on their participation in these conversations with quality content and timely dialectic input.

3. *Creating personal adventure, surprise and serendipity*: By playing to the museum's strengths in providing niche content and expertise, mobile social media has the potential to stimulate the wonder and the passion of the direct encounter with the physical museum for audiences using mobile platforms both on-site and beyond. Location-based technologies have the ability to put content and interpretation in the world in surprising and serendipitous places, giving the visitor a powerful experience linked to the museum despite the absence of the actual artifact. Augmented reality – overlays of text, images, or video on the real world seen through a mobile

device's camera – can bring not just information but haunting memories and emotions to everyday scenes; see, for example, the Museum of London's *Street Museum* app, which overlays the streets of London with images of those same locations from the 18th century and beyond at their correct geospatial coordinates, so that visitors can literally stumble upon the Great Fire of 1861 at London Bridge, suffragettes on Downing Street forty years later, Holborn Circus during the Blitz, or hippies in Hyde Park in 1970. Mobile social media referral and review services such as Foursquare and Gowalla incorporate gaming elements that can be adapted by the museum to create cultural tours and routes of discovery, with badges and other virtual rewards for fulfilling tasks. But for many, the simple pleasure of encountering others with similar interests can be enough to inspire enriching exchanges both among visitors and with the museum on the model of the best Twitter list.

4. Connecting the nodes: In the Museum as Distributed Network, every platform is a community, not just a point for content publication and distribution. By combining established social media platforms like Flickr, YouTube, Facebook, and Twitter with mobile social media, analog content, and publications built by or for the museum, we can create a whole that is greater than the sum of its parts. By thinking holistically about the museum as a network through its content rather than through the technology, we also avoid building orphans or one-offs that are difficult and even impossible to maintain as platforms evolve or become obsolete over time. Above all, iterative development is key to sustaining the content and the conversation over time, allowing them to grow and change with the museum's audiences as well as new scholarship and acquisitions.

Instead of simply publishing the same content in multiple places as in the multi-platform model, museums can adopt a distributed network approach to cross-platform design for a more sustainable model of building mobile experiences that are also more responsive, meaningful, and relevant to different audiences and their interests. Employing the principle that the best way to learn is to teach, design based on the distributed network model turns visitors into docents and ambassadors for our museums by facilitating and fostering communities of interest around our content and collections. The un-tour of the rhizomic museum takes the mobile experience from headphones to microphones, "from we do the talking to we help you do the talking." (Anderson 2009)

Epilogue: investing in the longtail

Since the era of the cabinet of curiosities, museums have been sanctuaries for specialists and their niche interests. As Chris Anderson has pointed out, it's often our niche interests or hobbies that elicit our deepest commitments and investment of time and energy. The museum is ideally positioned to tap into the power and passion of the longtail by connecting communities of interest through mobile social media and other means. In his keynote address at the 2009 *Smithsonian 2.0* conference, Anderson encouraged the Institution – and, by implication, all museums – not to be afraid to cultivate specialist content, ideas, and initiatives in quantity. Though many will fail, some will take root, flourish, and advance our missions (Anderson, 2009).

But resources are limited even at the best-funded of museums; where should we prioritize the investment of our hard-earned resources? The answer will obviously vary depending on each organization's individual circumstances, but here are my top ten tips for building a longtail mobile social media program that will yield benefits for both the visitor and the museum for the long term.

1. Invest in research and evaluation first. Even if you can't commission a bespoke study for your program or audience, make the time to consult colleagues and read and learn from the studies of others before beginning the planning process. Although we should be encouraged to innovate, to take risks, and to fail, there is no excuse for wasting money by reinventing the wheel – particularly when that wheel has already been demonstrated to be broken!
2. Mine existing assets. Even if you can't produce audio and video content for your mobile program, can you reuse existing text from labels and publications or the collections database? In combination with basic functionality like search, this simple content can provide valuable tools for the visitor to experience your collection more deeply and broadly.
3. Prioritize good writing. Even if you can't afford a new script and are forced to reuse wall labels,

catalog essays, and other existing assets, invest in an experienced writer who can make those texts speak fluently. A curator may be a great author, but writing for the ear is a very different skill set from writing for print.

4. Recruit good interlocutors. Quality may be in the eye of the beholder as Chris Anderson has argued (Anderson, 2009), but it's also important to recognize that in a polyvocal context, the museum's role and value is as a reliable and respected authority; therefore museum-authored content should be held to a higher standard than, say, user-generated content. Museum content does not have to be expensive, but it does have to be good; a *karaoke audio tour* can undermine trust in the museum as a source of quality, reliable content.
5. Cultivate diverse voices and points-of-view through user-generated content. A number of museums (including Tate Modern in 2003 – see Wilson, 2004; and the Powerhouse Museum – see Chan, 2009) have demonstrated the value of visitor input to their interpretive programs and collections information, both online and in the galleries. Not only can a plurality of voices add interest and new ways-in for visitors, but leveraging passionate amateurs and citizen curators can add real scholarship to the interpretive presentation as well.
6. Wherever possible, capture content in video (and HD and 3D...) The internet is increasingly a video platform

and even where video is inappropriate or distracting, its audio track and/or still images can be reused as standalone assets in other contexts. Years ago, Peter Samis issued a call for museums to video visiting artists, for example, even if there was no immediate use for them planned (Samis, 2005); those who heeded the call, like SFMOMA, are now sitting on a goldmine of interpretive assets. High definition images and video are quickly becoming the standard, and 3D is close within reach of museum budgets (See, for example, *The Mourners* online 3D sculptural exhibition by FRAME.[7]

7. Invest in low-cost recording devices, some training in basic editing, and iterate. There are many free apps and inexpensive audio and video recorders now that enable museum staff to produce content without requiring professional editing skills and facilities. Use them for rapid prototyping, test new content approaches with staff and visitors, and repeat! As Ted Forbes from the Dallas Museum of Art has said, "Do what you can, do it now, and do it better next time."
8. If you can't do anything else, podcast in iTunes U and iTunes. Podcasting is a classic longtail platform, building audiences slowly but steadily over time through push content. Anecdotally, several museums have reported more downloads of their podcasts and PDFs from iTunes U than even iTunes. Both are free to museums and offer either a paid or free distribution

channel to the public (downloads are free from iTunes U). Once you've put your downloadable content online, market it: let people know as soon as they visit your website that they can download your interpretation and bring it to the museum or enjoy it elsewhere.

9. If you can do a little more, build a mobile version of your website – or at least the pages dedicated to the visit and your interpretive program. Titus Bicknell has described how a mobile tour can be built in WordPress in as little as an hour (Bicknell, 2009). Several individuals and institutions have also made their mobile platforms available as open-source to the community (Alexander, 2008: Forbes, 2009; Bachta, 2009). HTML5 promises to bring dynamic mobile web development, with functionality that is currently only available within native apps, within easy reach of museum web teams.
10. Share what you've learned with your colleagues. One of the privileges of working in museums is that our work aims to make the world a better place. By sharing our knowledge and expertise generously, our value accrues exponentially across the industry as visitors satisfied with one museum experience are more likely to want to visit other museums, either online or in person, and help enhance the reputation of museums in general virally through their activities and personal contacts.

Notes

1. http://geschiedenis.vpro.nl/themasites/mediaplayer/index.jsp?media=19799217&refer nr=19265092&portalnr=4158511&hostname=geschiedenis&mediatype=video&portalid= geschiedenis ; translation by Ronden, 2010.

2. http://www.internetworldstats.com/stats.htm

3. http://www.itu.int/ITU-D/ict/newslog/Mobile+Phone+Subscribers+Pass+4+Billion+M ark.aspx

4. http://www.pewinternet.org/Reports/2009/12-Wireless-Internet-Use.aspx?r=1

5. http://museummobile.info/archives/136

6. http://museummobile.info/archives/136

7. http://www.themourners.org

An earlier version of this paper was presented and published as part of the proceedings of the 2010 Museums and the Web Conference on 31 March and 17 April 2010 at http:// www.archimuse.com/mw2010/papers/proctor/proctor.html

References

Alexander, Chris. *San Jose Museum of Art iPod Touch tour case study and webcasts* at http://wiki.museummobile.info/museums-to-go/projects/san-jose-museum-of-art Consulted January 31, 2010.

Anderson, Chris. *The Smithsonian's Long Tail*, keynote presentation to the 2009 "Smithsonian 2.0" conference, webcast at http://smithsonian20.si.edu/schedule_webcast2.html Consulted January 31, 2010.

Baudrillard, Jean. 1988. Simulacra and Simulations, *Jean Baudrillard, Selected Writings*, ed. Mark Poster, Stanford: Stanford University Press.

Bicknell, Titus. *No Time Like the Present: Rapid Deployment with Open Source Technologies - At What Cost?* presentation to Handheld Conference Online: From Audiotours to iPhones June 3, 2009 http://www.handheldconference.org/program/ Consulted January 31, 2010.

Chan, Sebastian. *Another OPAC discovery – the Gambey dip circle (or the value of minimal tombstone data)* on fresh + new(er): discussion of issues around digital media and museums, April 27, 2009 http://www.powerhousemuseum.com/dmsblog/index.php/2009/04/27/another-opac-discovery-the-gambey-dip-circle-and-the-value-of-minimal-tombstone-data/ Consulted January 31, 2010.

Finkelstein, Jonathan. 2009. *Unpublished interview with Len Steinbach and the author*, New York City, January 1..

Forbes, Ted. *Opensource tour solution webcast* on May 10, 2009 http://wiki.museummobile.info/museums-to-go/projects/

dallas-museum-of-art Consulted January 31, 2010.

Gescheidenis (History), "Draadloze rondleiding in het Amsterdamse Stedelijke Museum," film clip from *Polygoon Hollands Nieuws*, 28 July, 1952. Details at http://geschiedenis.vpro.nl/artikelen/19265092. Consulted January 31, 2010.

Haley Goldman, Kate. 2010. *Evaluation Report on Luce Foundation Center Mobile Content by the Fall 2009 Evaluation Theory & Techniques Class of Johns Hopkins University*, unpublished report.

Harris, Beth. Conversation with the author at Museums and the Web, Indianapolis, 2009; first reported in Nancy Proctor, *Evaluation-led Mobile Experience Design* January 3, 2010 http://museummobile.info/archives/297 Consulted January 31, 2010.

Johnson, T., D. Mitroff, and P. Samis. *Of Ansel and Atomz: Surfacing Deep Content On-line and On-Site at SFMOMA*, ed. J. Trant and D. Bearman, *Museums and the Web 2005: Proceedings*, Toronto: Archives & Museum Informatics, published March 31, 2005 at http://www.archimuse.com/mw2005/papers/johnson/johnson.html - I recall hearing Samis argue for the value of video in his presentation of this paper, but SFMOMA had in fact begun videoing artists many years before. Consulted January 31, 2010.

Griffiths, José-Marie, and Donald W. King, *Interconnections: The IMLS National Study on the Use of Libraries, Museums and the Internet*, School of Information and Library Science, University of North Carolina at Chapel Hill. The study was published as a series of reports over several years, beginning in 2006 on the Web. See www.interconnectionsreport.org.

Leggatt, Helen. "Gartner: Not long before mobile Internet access exceeds access via PCs," *Biz Report* January 14, 2010. http://www.bizreport.com/2010/01/gartner_not_long_before_mobile_internet_access_exceeds_acces.html Consulted January 31, 2010.

Longo, Eric. *Email to the author*, January 30, 2010.

Mann, Laura. Conversation with the author at Museums and the Web, Indianapolis, 2009; first reported in Nancy Proctor, *Manifesto for a New Mobile Architecture*, Museum Mobile wiki April 25, 2009. http://wiki.museummobile.info/archives/62 Consulted January 31, 2010.

Moad, Charlie. *TourML (in progress)* May 14, 2010. http://wiki.museummobile.info/museums-to-go/products-services/tourml Consulted January 31, 2010.

Petrie, M. and L. Tallon. The Iphone Effect? Comparing Visitors' and Museum Professionals' Evolving Expectations of Mobile Interpretation Tools. In *Museums and the Web 2010: Proceedings*, ed. J. Trant and D. Bearman. Toronto: Archives & Museum Informatics. Published March 31, 2010. Consulted April 21, 2010.

Proctor, Nancy. *Off Base or On Target? Pros and Cons of Wireless and Location-Aware Applications in the Museum*, paper presented at ICHIM 2005 http://www.archimuse.com/publishing/ichim_05.html/Proctor.PDF Consulted January 31, 2010.

Ronden, Vincent. Translation into English of Gescheidenis (History), *Draadloze rondleiding in het Amsterdamse*

Stedelijke Museum, film clip from Polygoon Hollands Nieuws, 28 July, 1952 at http://museummobile.info/archives/297#comment-5267 4 January, 2010. Consulted January 31, 2010.

Samis, Peter. Gaining Traction in the Vaseline: Visitor Response to a Multi-Track Interpretation Design for Matthew Barney: DRAWING RESTRAINT, in *Museums and the Web 2007: Proceedings*, ed. J. Trant and D. Bearman. Toronto: Archives & Museum Informatics, published March 1, 2007 Consulted January 31, 2010. http://www.archimuse.com/mw2007/papers/samis/samis.html

Samis, Peter. 2007. "Visual Velcro: Hooking the Visitor," in *The Digital Museum: A Think Guide*, ed. Herminia Din and Phyllis Hecht. American Association of Museums.

Smith, Koven, The Future of Mobile Interpretation. In *Museums and the Web 2009: Proceedings*, ed. J. Trant and D. Bearman. Toronto: Archives & Museum Informatics. Published March 31, 2009. Consulted January 31, 2010. http://www.archimuse.com/mw2009/papers/smith/smith.html

Whitney Museum of American Art: *Audio Guide Technologies Survey Final Report 2009* http://tatehandheldconference.pbworks.com/f/Whitney+Final+Report-+Appendices+G-I-revised.pdf Consulted January 31, 2010.

Wilson, Gillian. 2004. "Multimedia Tour Programme at Tate Modern," in *Museums and the Web 2004: Proceedings*, ed. David Bearman and Jennifer Trant. Toronto: Archives & Museum Informatics. http://www.archimuse.com/mw2004/papers/wlson/wilson.html Consulted January 31, 2010.

Museum Education and Mobile for Museums: Conceptual Considerations Between Old Issues and New Challenges

THEO HUG

University of Innsbruck

In the rapidly changing world of the internet and mobile technology, theory and research frequently struggle to catch up with developments, interactions, and permutations in technology – and evolving social forms and practices. As to museum experiences, issues of education and mediation have been discussed throughout history. In the 20th century, debates on entertaining, sensual education, and the education of the senses emerged. More recently, issues of mediatization, mobility, and media-enhanced museum experiences have emerged.

With respect to practical issues, many promises have been made when introducing distributed education, networked learning, and distance learning. Over the past years, many museums have been offering their visitors mediated learning experiences that go beyond traditional modes of exhibition and audio guides – for example, by making use of widespread portable devices and providing podcasts, cellphone tours, and platform-specific applications. Are we moving on from e-learning promises to m-learning promises? Are there useful crossover concepts in a situation of fragmenting knowledge, formats, audiences, and even life?

In this paper, I discuss old issues of education and mediation as they relate to museum experiences and paradoxical educational structures. These issues are outlined and rethought in view of recent literacy debates and processes of mediatization and digitization. Furthermore, a broad understanding of mobile learning and the potential of microlearning are argued as relational crossover concepts.

In this context, exemplary distinctions of micro-, meso-, and macrolevels for museum learning are highlighted. After discussing a number critical issues and questions, the chapter concludes with an appeal for forms of education based on principles of bricolage.

Historical aspects and modern museum education

Museums have always been "educational" if we think of a broad understanding of education. This accounts for ancient Greek temples (mouseion) where inspiring kisses of caring young goddesses named Muses could be expected, as well as for modern art exhibitions and outreach activities of science centers. As a matter of fact, informal educational contexts and target audiences have changed throughout history, along with different understandings of what a museum is (cf. Alexander, 1996). Looking at European history, explicit didactic or educational claims were not made until the 16th century (cf. Vieregg, 1990; Parmentier 2009). As to museum didactics,[1] the beginnings of development can be confirmed with the Kunstkammer (1563–67) of the Wittelsbach Duke Albrecht V in Munich. The Duke's personal physician, Samuel van Quiccheberg, a collector, coined the idea of the ideal museum as a place which also serves the purpose of imparting practical knowledge in his treatise *Inscriptiones vel Tituli Theatri Amplissimi* (1565) (cf. Roth, 2000). Further developments occurred in the course of museum establishment in the nineteenth century. Schmeer-Sturm summarizes them as follows:

> *Didactic considerations informed the conceptions of several new museums, whether they were oriented more to representation and politics, like in the case of the sovereigns, or intended as national monuments. The Fürstenmuseen of the nineteenth century were conceived as educational institutions which made it possible for the bourgeoisie to become familiar with the blossoming of culture as a merit of the nobility. In contrast, some museums became established which were supported by bourgeois historical and museum societies and aimed at historical education and the cultivation of taste. 1826 marked the first time that school classes were given lessons in a museum (Frankfurt: Senckenberg Museum). (Schmeer-Sturm, 1997, p. 1093)*

Around 1900, Munich became the center of efforts related to museum education.[2] The professionalization of museum education on an international level and establishment of pertinent institutions started in the middle of the 20th century.[3]

As to mobility and portable units, various modalities such as stealing, borrowing (loan exhibitions), or peregrinating (traveling exhibitions, road shows) played a role. Mobility mostly referred to the learners when doing field trips and *learning walks* (Lerngänge). The conception is widespread to this day that museum education essentially consists of an educator leading a group of learners through the museum or to historic places. But in the course of the professionalization of museum education over the past 60 years, didactic models, training methods, and media programs have been enhanced and differentiated extensively. In his overview of museum education as exhibition didactics, Hey (1978) mentions eight points ranging from tours and programs for exhibition visitors to educational materials and resources, as well as classes and outreach activities. In a more recent study Wetterlund and Sayre (2009) distinguish seven fields[4] and altogether more than 55 different kinds of tasks and educational activities. Schmeer-Sturm (1997, p. 1100) explores the examination and preparation of museum objects in relation to the life horizon of the participants. She highlights four possibilities (ibid.):

- references to the life horizon of the visitor and his or her everyday life;
- references to the social system of the visitor;
- connections between the history of the visitor's home region and the visited region;
- connections between present-day life in the visited country and the visitor's own past or near future.

The basic idea is that meaningful educational experiences are more likely if the selection of references or connections is based on the knowledge, interests, age, family background, and social status of the participants or visitors.

New challenges?

Since the 1990s, an increasing presence of digital media has been observed in museums as well as in other institutions and social spaces. Many museum directors and curators have listened to up-and-coming e-promises such as more effective marketing strategies and increasing audiences, educational improvements and cost reduction, better contextual information on exhibits, or the democratization of knowledge. Undoubtedly, the scope of museum education technology has been expanded, and more and more fascinating digital mobile examples are in use. Once again, media-enhanced museum experiences were on the agenda (cf. Stogner, 2010). But has the scope of museum education been enhanced? What are the new challenges?

First of all, many known challenges did not disappear. For the majority of museums, marketing pressures, the need for promotional events, and the pursuit of higher income are serious issues not only with regard to educational tasks, but also in terms of general functions such as collection, conservation, research, or social action. Furthermore, there are known tensions, antinomies, and dilemmas which played and still play a role in museum education. Here are some examples:

- scientific quality and accuracy versus popular understanding and public accessibility;
- museums in media, media in museums, and museums as media and mediated institutions;
- expectations and requirements of preservation and change;
- increasing pressure for decisions and increased demands for legitimation;
- professional dominance and risk of collapse;

- identifying arrangements for learning and leaving spaces open for future developments;
- education and/or entertainment;
- promises of learning experiences and the structural uncertainty of learning outcomes;
- formal structures and traditional decisions versus creative organizational developments;
- institutional autonomy and interdependency;
- sharing knowledge and social obligations versus fierce competition and shaping profiles;
- social nearness and professional distance.

The list could be extended. But even this short version shows that paradoxical structures are of general relevance. Corresponding challenges vary in relation to historical constellations as well as to socio-cultural, economic, cognitive, and affective dynamics. Accordingly, there are many solutions; there are no final answers, but rather arrangements with what has been called "ultimateness of tentativeness" (cf. Schmidt, 2010).

One way of understanding the new qualities of challenges is to consider wider contexts and their relevance for museum education and respective claims. If we look at social descriptions and time diagnosis today, some of them are bound to the postmodern condition in terms of post-industrial, computerized societies. Others focus on different aspects, which are expressed by concepts such as: multicultural society, multi-option society, world society, education society, adventure society, fun society, communication society, information society, knowledge society, or media society. Each of the descriptions opens up a particular horizon of discourse and analysis, and even if we are aware of their limits we cannot focus on the blind spots of the concepts while using them.

With this in mind, it is apparent that many decision makers become uneasy with clear indications for the future of education – especially since voices question pedagogical claims and processes of "pedagogization" and "didactization" and even museum education itself (cf. Gruschka, 2002, pp. 366-376). The reflection of meta-processes remains crucial, regardless of whether we conceptualize paradoxes and ambivalences – such as the high value of economic performance versus missing justice of chances, the ongoing commercialization of knowledge versus knowledge obligation to the community, or others mentioned above – in terms of reflexive modernization, postmodernity and difference, or even *After Society* (Faßler, 2009). Looking at current developments, media appear to be important driving factors, not only as media technologies or media institutions, but rather in the sense of complex interplays between cultural, technological, and social dynamics (cf. Rusch, 2007).

As a consequence, we are well advised to conceptualize historical constellations as mediated and mediatized historical constellations. Having said this, mediation means more than an aspect of technology which is opening up new spaces of communication and affecting habits, value systems, and world views or changing modes of perception and thinking. It means continual transformation of the human condition in all its physical, psychological, biological, socio-cultural, and economic aspects. While mediation as a rather broad term refers to all sorts of different phenomena – "everything is mediated" (cf. Livingstone, 2009) – the term *mediatization* refers to "the meta process by which everyday practices and social relations are historically shaped by mediating technologies and media organizations" (Livingstone, 2009, p. 2; see also: Lundby, 2009): a process which results in ongoing and increasing dynamics of media communication on temporal, spatial, and social levels. In line with these considerations, mediality can be characterized as follows:

> *By their activities in media systems men create media-worlds (or media-realities) which compete with one another. Media systems work as observing and describing systems which do not start from 'the reality' but from former descriptions of reality which accordingly are transformed into or followed by new descriptions. On this note, the description of reality and the reality of description coincide. This argument resolves the tedious question regarding the relation of media and reality. Due to their system-specific logic media cannot represent an extra medial reality in an objective way. Instead, they can only produce and present medium-specific realities. (Schmidt, 2008, p. 97).*

Moreover, there is no withdrawal of communicativity and

mediality, because:

> *Discourses function via the co-presence of materiality and meaning construction processes. This contemporarity defines the mediality of our relation to the world. Language is inseparably bound to materialities; media are necessarily bound to technicality. (Schmidt, 2008, p. 101).*

Media technologies, media institutions, and communication instruments such as languages have both expanded and disciplined our forms of perception and action spaces in relation to medium-specific historical conditions and constellations. This explains "why there are literates and illiterates for every medium" (Schmidt, 2008, p. 96) and why new challenges are consistently related to new media developments and respective mediated constellations.

As a matter of fact, this accounts for ongoing discussions about the changing definitions of roles and identities of museums; their functions as community spaces, gathering places, and public spheres; policy development and shifts in power and authority; as well as new models for learning, education, and collaboration (cf. Pastore, 2009). And as I understand it, many (post-) modern concepts and practices of museum education are responding to this situation by offering a plurality of narrations, mostly voting for constructivist and learner- or audience-centered approaches (cf. Roberts, 1997; Hein, 1998; Hooper-Greenhill, 1999; Falk & Dierking, 2000; John & Dauschek, 2008; Simon 2010).

As to the question of literacies, I want to distinguish

between two important aspects. First, there is the ongoing discussion on information literacy, media literacy, and computer literacy, and their relation to traditional literacies and associated fields like visual competency, digital fluency, or innovation skills. Moreover, other literacies are in debate, such as numeric literacy, environmental literacy, and various "new literacies". As to museum and library development, the Institute of Museum and Library Services (2009) provides a rich and useful synopsis with *Museums, Libraries, and 21st Century Skills*. Sympathetic to ecological approaches, but with a more general intent, Henry Jenkins et al.(2006) list the following "new literacies" in their white paper *Challenges of Participatory Culture: Media Education for the 21st Century*:

- **Play** – the capacity to experiment with one's surroundings as a form of problem-solving.
- **Performance** – the ability to adopt alternative identities for the purpose of improvisation and discovery.
- **Simulation** – the ability to interpret and construct dynamic models of real-world processes.
- **Appropriation** – the ability to meaningfully sample and remix media content.
- **Multitasking** – the ability to scan one's environment and shift focus as needed to salient details.
- **Distributed Cognition** – the ability to interact meaningfully with tools that expand mental capacities.
- **Collective Intelligence** – the ability to pool

knowledge and compare notes with others toward a common goal.

- **Judgment** – the ability to evaluate the reliability and credibility of different information sources.
- **Transmedia Navigation** – the ability to follow the flow of stories and information across multiple modalities.
- **Networking** – the ability to search for, synthesize, and disseminate information.
- **Negotiation** – the ability to travel across diverse communities, discerning and respecting multiple perspectives, and grasping and following alternative norms." (Jenkins et al., 2006, p. 4; emphasis in the original)

Related discussions on these and other (new) literacies are important and necessary in view of current and foreseeable media constellations. They convey academic and public discourse and they can act as aids for planning and decision making.

The second aspect refers to widespread claims that these skills "build on the foundation of traditional literacy, research skills, technical skills, and critical analysis skills taught in the classroom" (Jenkins et al., 2006, p. 4). On the one hand, building on traditional literacies is a necessity, and not a question. On the other hand, when we speak of various (new) literacies in taken-for-granted ways, embedded tensions are easily underestimated. For the most part, traditional literacy is based on the primacy of written language focusing on continuity of forms, meanings, and semantic systems. Visual thinking, for example, as far as it is mentioned at all, is meant rather for the subject of arts and not for science and reliable epistemology. In other words, the traditional significance of alphabetical and print literacy is biased; or as McLuhan puts it, schools are "custodians of print culture" (1962, p. 215), at least partly aiming to protect print-based cultures and shield new generations against threatening (new) media effects.

Since museums traditionally tend to value all the senses, museum education may be in a better position compared to formal education in schools. However, for both types of institutions, dealing with single literacies or technologies in isolation appears to be problematic. Having said that, taking ecological approaches is associated with a need for crossover concepts and development of meta-competencies in order to reorganize skills, literacies, and education.

Museum education, mobile learning and microlearning

There is no doubt that the scope of museum education technology has expanded. Apparently, more and more institutions are using mobile devices and mobile learning. Although the concept of "mobility" is anything but new, recent examples of digital mobile initiatives[5] show that new options for information flows and encouragement of learning and education have arisen – along with processes of digitization and various micro-developments. Various forms of mobile communication have been promoted due to the relative ubiquity, portability, and multi-functionality of mobile devices and the increasing combination of functions in a single device, as well as ongoing convergence of media, cultures, and technical services (cf. Nyíri, 2003a; Katz, 2008); chances for learning have also been promoted across institutional boundaries, different action spaces, and formal and informal contexts.

Context-crossing dynamics play important roles not only with regard to performative aspects of acting or moving but also to cognitive skills like knowing, thinking, or remembering. They have to be considered on conceptual levels as well. A promising way of consideration is based on ecological thinking in terms of dynamic relations of individuals and contexts, nested levels, and multiple dimensions of temporality. Inspired by this way of thinking, the basic distinction of micro-, meso-, and macrolevel has been applied in many fields, subjects, and disciplines. One area is learning and as a result, the learning of individuals, organizations, generations, or societies can be distinguished, as well as micro-, meso-, and macrolearning (see Fig. 1 below).

As microlearning has developed, it can be said to have potential and factual usefulness as a relational crossover concept (cf. Hug, 2007). Microlearning is a rather new expression, in contrast to microteaching and also to other concepts such as microphysics or microbiology. Similar to related expressions like microcontent, micropayment, or micromedia, microlearning has only been in use since about 2002 – though many aspects of learning, didactics, and education have been addressed on what can be called a "micro-level" for centuries (cf. Hierdeis, 2007).

The discourse, or rather, multiple discourses that have emerged and developed around microlearning, is above all polyvocal and international. Such cultural, geographic, and linguistic heterogeneity emphasizes that the technology central to microlearning – like any technology or technical system – is not constituted in isolation, but produces the same results in different institutional, social, and cultural contexts.

The heterogeneity of the contexts, cultures, and meanings associated with microlearning is further augmented by the heterogeneity of the term *learning* itself. For example, learning can be conceptualized as a process of building up and organizing knowledge. But learning can also refer to the change of behavior, attitudes, values, mental abilities, task performance, cognitive structures, emotional reactions, action patterns, or social dynamics. Moreover, if we think of learning as processes of transformation which are based on activities of meaning-making or behavior modification, we may think, for example, of:

- cognitive, emotional, body, or social learning;
- learning driven by motivation, technology, problem, market, or interest;
- goal oriented, problem based, situated, or meaningful learning;
- instrumental/mechanistic or expansive learning;

- functional or self-reflective learning;
- living or alienated learning;
- conscious or unconscious dimensions;
- self-organized and externally organized learning – learning learning;
- learning of individuals, organizations, generations, or societies;
- formal or informal contexts;
- and forms of learning learning.

From a meta-perspective, numerous modes of describing relations in terms of micro-, meso-, and macrodimensions can be distinguished. It depends on whether the frameworks and application contexts are useful or not. So, for example, the renowned eco-systemic model of Urie Bronfenbrenner (1986, 2005) describes one helpful mode of distinguishing between micro (relations of humans to other humans or groups), meso (relations of a human as a whole), and macro dimensions (totality of relations in a society). This applies to microlearning issues, too. *Microlearning* can be defined as characterizing content, processes, technologies, competencies, or learner groups. The key to microlearning is that which occurs at the most minute of levels, as opposed to the meso- or macro-: minutes or seconds of time are relevant instead of hours, days, or months; sentences, headlines, or clips are the focus rather than paragraphs, articles, programs, or presentations; and portable technologies and loosely-coupled distributed environments are of interest rather than monolithic or integrated turnkey systems.

The selection of exemplary distinctions of micro-, meso-, and macrolearning in Fig. 1 shows how these conceptual considerations can be applied in the context of museum learning .

Regarding the variety of microlearning concepts, their potential and factual usefulness can be pointed out by the example of mobile learning. Why are mobiles important for microlearning? There are several reasons. Firstly, content depicted on mobiles typically is microcontent. Second, attention spans and periods of time are in many cases relatively short. Third, mobile screen sizes are typically smaller than other screen sizes. Fourth, there are a great variety of possibilities for creating and integrating micro-steps in comprehensive learning processes, including learning in informal contexts. Fifth, microlearning is in many ways related to physical, social, and cognitive mobility, as in *Moving Cultures* (cf. Caron & Caronia, 2007). Sixth, mobile devices are often related to micro-platforms and micromedia business models. The list of arguments could be continued and argued in detail. Yet, these six points provide plausible reasons for the importance of mobiles in microlearning.

When thinking of the concepts, applications, and practices of mobile learning and mobile communication, we have to be aware that technology-driven considerations are of limited use both generally and in the context of museum education. If we want to understand the interplay of physical, social, and cognitive mobility, we have to look at combinations of

Museum learning: examples of the distinctions between micro-, meso-, and macro-levels				
	Content	Meaning-making	Competency	Technology
Micro level	Learning objects, single parts of a collection are being considered for a special learning purpose.	Individualized learning, visitors construct their own meaning according to socio-cultural backgrounds.	Special competencies of learners, teachers and museum staff.	Focusing on single devices and screen-sized media offers.
Meso level	Learning experiences based on collections.	Group learning based on audience-institution collaboration, organizational learning .	Museum as a competence center.	Integration of community stakeholder inputs when planning for educational technologies.
Macro level	Learning experiences based on intercultural programs or exchange of collections.	Meaning-making as learning of generations, cultures and of societies.	Audiences and partner institutions are involved in collaborative planning for competence development.	Inter-institutional strategic planning for learning technologies, collaborative use of learning technologies to foster programs.

Figure 1: Microlearning – mesolearning – macrolearning: a selection of exemplary distinctions.

technology- and user-related factors.[6] Accordingly, mobile learning:

> *is not about delivering content to mobile devices but, instead, about the processes of coming to know and being able to operate successfully in, and across, new and ever changing contexts and learning spaces. And, it is about understanding and knowing how to utilise our everyday life-worlds as learning spaces. (Pachler et al., 2010, p. 6)*

In addition, this notion of "coming to know" should be expanded according to the differentiations mentioned above, especially with regard to emotional, bodily, and unconscious dimensions – as well as distinctions like functional and self-reflective learning or instrumental and expansive learning.

As to the didactics of mobile learning and microlearning, it is important to be aware of different cultural and academic traditions. It makes a big difference if we understand didactics mainly as instructional design, as literary genre, as it is discussed in the French speaking cultural area – or as an umbrella term which refers to a variety of concepts, approaches, models, theories, experiences, technologies, or questions of the art of teaching and learning.[7] However, didactic considerations may focus on subjects (who); contents and skills (what); methods and technologies (how); reasons, purposes, and goals (why and what for); as well as on social relations, institutional and social conditions, settings and arrangements, learning ecologies and cultures, media environments, power and control, or evaluation and assessment.

As mentioned above, the term *microlearning* was not used until a decade ago. Nevertheless, related phenomena have always played a role in institutional learning contexts in general and in museum education in particular. The question is rather, how micro steps and short-term learning activities are positioned, situated, structured, contextualized, valued, combined, complemented, or contrasted – and which forces are at work in institutional contexts and elsewhere. Generally speaking, there are many ways in which the association of microlearning aspects can be conceptualized and linked to comprehensive models, besides incidental and unavoidable modes of punctuating and highlighting single learning steps in workflows and everyday life.

To some extent, we have to rethink synthetic and analytic methods of dealing with small steps and structuring them (or not). Moreover, the processes of mediation and digitization provide reasons to rethink didactics. Microlearning "radicalizes a development that began with the introduction of books and images to the school" (Hierdeis, 2007, p. 38); microlearning "creates loopholes in the learning processes in educational institutions, since the acquisition of information can occur without pre-established aims, without outside control and evaluation" (ibid.).

If we accept this need to rethink traditional models and enhance didactical thinking, the question arises of how this can be done. I think this can be done by means of analogical thinking and figurative transmission of applicable models that are used in other fields. There are quite a few general models that introduce different and viable ways to understand how microlearning elements can be linked together. Here are three examples (cf. Hug, 2007, p. 20):

- In the aggregation model microlearning elements that are fundamentally similar are bundled or combined as a relatively unstructured entity or homogenous mass ("aggregate").
- In the conglomerate model diverse microelements are arrayed as a kind of assortment or "bouquet" of learning products and processes.
- In the emergence model new phenomena, coherent structures and qualities evolve from and between microlearning elements themselves. These novel patterns or properties cannot be attributed to any single element. Instead, they arise out of a multiplicity of relatively simple interactions or steps in the dynamic process of self-organization.

In looking at traditional learning arrangements and design patterns, models like the four mentioned offer viable ways for renewal. Such models may have potential for the substitution of traditional didactical orientations, settings, and concepts as a matter of principle; theoretically and practically making use of them might work better in complementary ways.

According to the (new) ways of modeling, the form of final microlearning results may have characteristics of fragments, facets, episodes, skill elements, or discrete tasks. Even though we have manifold options for describing and creating related

	Education as the competence of plurality	**Education as reflective ability to learn**	**Education as *antagonistic education***
Explanation	Transversal competence, qualification to sensibly deal with and intervene in plural situations.	Qualification to reflect processes of learning from several perspectives.	Overcoming discursive constraints, abdication of harmonizing ideals.
Thematic relevance	Dealing with lifeworlds as mediated worlds, metaphorical competence.	(Re-)organization of learning processes, media, effect and attention management.	Reference modalities, modularization and modalization of the experience of reality.
therefore	Education as media education – education with media instead of against them. Education as the ability and qualification to manage differences.		

Figure 2: Education based on principles of *bricolage* (cf. Hug, 2008a, p. 64).

phenomena, microlearning's general characteristics are easily outlined: in terms of time, microlearning is related to relatively short efforts and low degrees of time consumption, and in terms of content, it deals with small or very small units and rather narrow topics. In other words, microlearning works as a crossover concept in the sense of special moments or episodes of learning while dealing with specific tasks or content, and engaging in small but conscious steps.

With these characteristics in mind, the understanding of education as *bricolage* (see Fig. 2 above) appears to be an appropriate contemporary conception of education.

In contrast to modern, realist, and truth-oriented

understandings, education as bricolage favors postmodern, constructivist, and contextualist orientations. In the sense of Lévi-Strauss (1968), principles of bricolage have been applied to educational contexts in various ways – principles such as openness, agility and flexibility in thinking and acting, dealing with heterogeneous matters and limited resources, and spatio-temporal anchoring of actions, (cf. Schön, 1991). Of course, one can again argue that these principles have played a role throughout the history of education. But after postmodern discourse emerged, the principles have been reappraised and appreciated more clearly (cf. Turkle and Papert, 1990).

A similar development can be described from strong and holistic reasoning toward weaker forms of reasoning (cf. Welsch, 2000). It was Jean-François Lyotard (1989) who pointed out philosophical problems under plural conditions most profoundly. The term *antagonistic education* refers to his considerations on language games in conflict and *The Differend* (1989); they can be used as fruitful points of departure and sources of inspiration in the context of the development of educational concepts and theories. Moreover, they remind us to be aware of the importance of transitional processes and dynamics at the interfaces of unrelated discourses. This explains the focus on competence of plurality and the role of both transversal competencies (cf. Perger and Hug 2000) and learning cultures of deliberation (Blanck, 2002) in plural and fragmented lifeworlds.

Critical issues and further questions

The conceptual considerations discussed are based on interrelationships between cultural and technological developments and an awareness of differences, ongoing transitions of media and learning cultures, and a manifold "simultaneity of the non-simultaneous" (J. Habermas). Regardless of whether we interpret them in eco-systemic or in post-ecologist terms, holistic educational ideals are evaluated with them. Nevertheless, mobile learning, microlearning, and education as bricolage can correspond to meaningful didactical settings both for educators and learners. But are museum educators willing to accept the "postmodern" challenges (cf. Hooper-Greenhill, 1999, p. 71), to address the needs of diverse audiences, and above all to take a proactive role in a "world marked by fluidity, provisionality and instability, where responsibilities for meaning-making as well as risk-taking are firmly located with the individual" (Pachler et al., 2010, p. 11)?

Even if we think of microlearning and bricolage as fruitful and useful concepts, however, critical issues and questions remain, such as the following:

- Aren't the pursuit of the "real thing" and the promotion of media-enhanced museum experiences more about high-tech gadgetry, "Disneyfication" (cf. Roberts, 1997, p. 69), and "flickering minds" (Oppenheimer, 2003) than about museum education? Isn't microlearning, the problem, considered itself to be the solution?
- Are developments of micromedia, microcontent,

and microlearning more than temporary hypes? Are we moving on from empty e-learning promises to m-learning promises? Isn't it all about loosely joined bits and pieces (D. Weinberger), at the same time underestimating systemic dimensions and power relations?

- Isn't "postmodern" museum education just another problematic version of "education towards truth" (cf. Hug, 2008b)? Aren't the concepts of de-governmentalization turning out to be concepts of re-governmentalization on other levels (ibid., p. 251)?
- Isn't microlearning mainly about computer literacy or digital fluency? Which literacy modes have to be considered if we want to avoid technology fallacies?
- In looking at contemporary societies in flux and shifting socio-cultural, political, and economic structures: How can museum educators meet the needs of diverse and mobile audiences in view of their limited resources and scopes of action?

In many places, technology-enhanced learning has promoted bureaucracy and spheres of influence of ICT-administrators rather than innovation in education. Evidence for related tendencies of "techno-bureaucracy" (Varis, 2006, p. 2) can easily be found in indicators such as: the primacy of controlling in relation to didactical innovation, forcing educationalists to spend hours on reports for paper-shufflers, and lavishing money on technology while passing off educational efforts for peanuts. In many countries and educational institutions, issues of administration, accounting, and controlling are too centric – even though media developments and learner interests are creating a situation in which new learning cultures and more flexible educational alternatives are needed. Mobile learning and microlearning approaches can promote such alternatives; however, they depend on successful solutions with regard to personal, institutional, and medial aspects of learning environments.

Conclusion

In concluding, I want to emphasize three aspects which I think especially important for future museum education. Unsurprisingly, they are situated in cross-cutting areas between old issues and new challenges. And above all, it is unsurprising that throughout history – from Greek temples and treasure chambers to cabinets of curiosities (*Kuriositätenkabinette*) and *Wunderkammern* to the modern museums and *Travels in Hyperreality* (Eco, 1986) – some themes recur in different versions. However, if we are going to conceptualize museum education in future-oriented ways, considering the following aspects seems to be crucial in view of moving cultures and fragmenting knowledge, formats, and audiences:

- The question of how to (re)organize the past suggests itself. Who counts as legitimized speaker or storyteller in which context and how are authorities established? There is the task to balance social and individual contexts of remembering, allowing for the reflection of chronology, geography, and politics of memory (cf. Kumar et al. 2006). Moreover, museum education is one of the most sensitive fields where the didactics of memorization (cf. Schäfer 2009) and politics of remembering come together.

- The (traditional) significance of alphabetical and print literacy must be rethought and re-conceptualized in relation to literacy concepts and "new" literacies. In doing so, we have to be aware of limits of the transmission and application of literacy concepts to various contexts in which not everything is about writtenness or alphabetical order. On the contrary, relational concepts of literacy, numeracy, and picturacy are suggested by aspects such as: ductility (*Bildsamkeit*), visual culture, spatial and temporal orientation, ability to act appropriate to situations and contexts, and forms of abstraction.
- Last but not least, relations between education and edutainment need clarification depending on contexts and goals. Some museum educators seem to take edutainment for granted without worrying about concerns of McDonaldization or Disneyfication. On the other hand, some critical educationalists think of edutainment as "mission impossible" because of their strictly discrete concepts of education (*Bildung*) and entertainment. In my view, education and entertainment are not incompatible; for example,

> Friedrich Schleiermacher (1768-1834) included conviviality as an important aspect in his conception of *Bildung* 200 years ago. Bridging can be done by balancing and communicating different perspectives. This applies to investigating limits for the use of game-based approaches and playful elements, in the context of intentional didactical purposes and political goals of democracy, as well.

For these and other aspects – like, for example, museum education as empowerment, experience, ethics, or narrative (cf. Roberts, 1997) – communicative processes of stabilization and de-stabilization of educational formats and claims are essential. In this sense, the approaches suggested here are not strictly pedagogical in terms of traditional regulations. Rather, museum exhibitions may become educational without using pedagogical methods, and pedagogical interventions may "work" without becoming educational. As to questions "concerning the senses, their training through educational and mediatic forms correspondingly, will only increase in their urgency and diversity" (Friesen, 2010). The potential of museums for advancements in "sensuous education" should not be underestimated. There are chances for fruitful perspectives on museum education and bridging – popular culture and school environments, lifeworld and technology, as well as learning in informal and formal contexts – as long as we do not mix up museum education and museum education technologies, and as long as we are aware of relevant paradoxical structures.

Notes

1. I use the term 'didactics' as an umbrella term in the wide sense of theory and practice of learning and teaching as it is widely used throughout Europe and especially in German-speaking countries. Corresponding understandings are related to Bildung and educational philosophy rather than to narrower understandings in the sense of instructional principles or design (cf. Friesen, 2007).

2. This is documented through the activities of Hugo von Tschudi (1851–1911), director of the Pinakotheken; Oskar von Miller (1855–1934), founder of the techno-historical Deutsches Museum; and Georg Kerschensteiner (1854–1932), who tried to incorporate the manual, technical and practical work as well as the engagement with museums into the activity school (Arbeitsschule), and last but not least, Adolf Reichwein (1898–1944), head of the "School and Museum" department in Berlin who developed a pedagogical concept for the working methods of teachers and students in museums (cf. Schmeer-Sturm, 1997, p. 1094).

3. Cf. International Council of Museums (ICOM, see http://icom.museum/) and Council of Education and Cultural Action (CECA, see http://ceca.icom.museum/).

4. Namely tours, informal gallery learning programs, community, adult and family programs, classes and other public programs, partnership with other organizations, school programs, and online educational programming (cf. Wetterlund & Sayre, 2009, pp. 36-42).

5. See, for example, Tate Multimedia Tour (http://www.tate.org.uk/modern/multimediatour), Electronic Guidebook Research Project (http://www.exploratorium.edu/guidebook/), Multimedia Mobile Museum – Museum of Marble in Carrara (http://www.medien.ifi.lmu.de/mirw2007/papers/MGuides2007_Santoro.pdf), Mobile for Museums – Center for History and New Media (CHNM) at George Mason University, Virginia (http://chnm.gmu.edu/labs/mobile-for-museums/).

6. In „Thinking About the 'm' in Mobile Learning" Kress and Pachler (2007, pp. 142-144) provide a number of starting points such as flexibility and portability, multi-function-

ality, multimodality, nonlinearity, interactivity, and communicative potentials for future research on mobile learning. For further conceptual starting points see also Nyíri (2003b).

7. In their pedagogical learning theory Göhlich and Zirfas (2007) distinguish between knowledge-learning, ability-learning, living-learning, and learning-learning. These distinctions could be further elaborated in terms of micro-, meso-, and macro-learning, too.

References

Alexander, Edward P. 1996. *Museums in Motion: An Introduction to the History and Functions of Museums*. Walnut Creek: AltaMira Press.

Blanck, Bettina. 2002. *Erwägungsorientierung, Entscheidung und Didaktik*. Stuttgart: Lucius & Lucius

Bronfenbrenner, Urie. 1986. Recent Advances in Research on the Ecology of Human Development. In *Development as Action in Context - Problem Behaviour and Normal Youth Development*, edited by Rainer K. Silbereisen, Klaus Eyferth, and Georg Rudinger, pp. 287–310. Berlin: Springer.

Bronfenbrenner, Urie, ed. 2005. *Making human beings human: bioecological perspectives on human development*. Thousand Oaks: Sage Publications.

Caron, André H. and Letizia Caronia. 2007. *Moving Cultures: Mobile Communication in Everyday Life*. Montreal/Quebec: McGill-Queen's University Press.

Eco, Umberto. *Travels in Hyperreality. Essays*. New York: Harcourt Brace and Co., 1986.

Falk, John H. and Lynn D. Dierking. 2000. *Learning from Museums: Visitor Experiences and the Making of Meaning*. Walnut Creek: AltaMira Press.

Faßler, Manfred. 2009. *Nach der Gesellschaft*. München: Fink.

Friesen, Norm. 2007. (Micro)Didactics: A Tale of two Traditions. In Didactics of Microlearning. Concepts, Discourses and Examples, edited by Theo Hug, 83-97. Münster u. a: Waxmann.

Friesen, Norm. 2010. Vision and "the training of perception":

McLuhan's Medienpädagogik. In *Visuelle Kompetenz*, edited by Theo Hug and Andreas Kriwak. Innsbruck: Innsbruck University Press (forthcoming).

Göhlich, Michael and Jörg Zirfas. 2007. *Lernen. Ein pädagogischer Grundbegriff.* Stuttgart: Kohlhammer.

Gruschka, Andreas. 2002. *Didaktik. Das Kreuz mit der Vermittlung. Elf Einsprüche gegen den didaktischen Betrieb.* Wetzlar: Büchse der Pandora.

Hein, George E. 1998. *Learning in the Museum.* London: Routledge.

Hein, George E. 1999. The constructivist museum. In *The Educational Role of the Museum*. edited by Eileen Hooper-Greenhill, 73-79. New York: Routledge

Hey, Bernd. 1978. *Die historische Exkursion. Zur Didaktik und Methodik des Besuchs historischer Stätten, Museen und Archive.* Stuttgart: Klett.

Hierdeis, Hemwart. 2007. From Meno to Microlearning: A Historical Survey. In *Didactics of Microlearning. Concepts, Discourses and Examples*, edited by Theo Hug, 35-52. Münster u. a.: Waxmann.

Hooper-Greenhill, Eilean. 1999. Museum learners as active postmodernists: contextualizing constructivism. In: *The Educational Role of the Museum*, edited by Eileen Hooper-Greenhill, 67-72. New York: Routledge.

Hug, Theo, ed. 2007. *Didactics of Microlearning. Concepts, Discourses and Examples.* Münster u. a.: Waxmann.

Hug, Theo. 2008a. Medienphilosophie und Bildungsphilosophie – Schnittstellenerkundungen. *In Philosophy of the*

Information Society. Proceedings of the 30. International Ludwig Wittgenstein Symposium Kirchberg am Wechsel, Austria 2007. Volume 2. Edited by Herbert Hrachovec and Alois Pichler, 43-73. Frankfurt u. a.: Ontos.

Hug, Theo. 2008b. Education towards Truth? Reflecting a Sentence of Josef Mitterer. In *The Non-dualizing Philosophy of Josef Mitterer. Constructivist Foundations*, vol. 3, no. 3, edited by Alex Riegler and Stefan Weber: 249-253,.

Institute of Museum and Library Services Museums, *Libraries, and 21st Century Skills* (IMLS-2009- NAI-01). Washington, D.C. (2009) Retrieved April 15, 2010, from: http://www.imls.gov/pdf/21stCenturySkills.pdf.

Jenkins, Henry et al. *Confronting the Challenges of Participatory Culture: Media Education for the 21st Century.* White paper co-written for the MacArthur Foundation. (2006). Retrieved April 15, 2010, from: http://www.projectnml.org/files/working/NMLWhitePaper.pdf

John, Hartmut and Anja Dauschek, eds. 2008.*Museen neu denken: Perspektiven der Kulturvermittlung und Zielgruppenarbeit*. Bielefeld: Transcript.

Katz, James E., ed. 2008. *Handbook of Mobile Communication Studies*. Cambridge: MIT Press.

Kress, Gunther and Norbert Pachler. 2007. Thinking About the 'm' in Mobile Learning. In: *Didactics of Microlearning. Concepts, Discourses and Examples*, edited by Theo Hug, 139-154. Münster u. a.: Waxmann.

Kumar, Keval J., Theo Hug, and Gebhard Rusch. 2006. Construction of Memory. In *News in Public Memory. An*

International Study of Media Memories Across Generations, edited by Ingrid Volkmer, 211-224. Frankfurt a. M. u. a.: Lang.

Lévi-Strauss, Claude. 1968. *The Savage Mind*. Chicago: University of Chicago Press.

Lyotard, Jean-François. 1989. *The Differend: Phrases in Dispute*. Minneapolis: University of Minnesota Press.

Nyíri, Kristof, ed. 2003a. *Mobile Communication. Essays on Cognition and Community*. Wien: Passagen.

Nyíri, Kristof, ed. 2003b. *Mobile Learning. Essays on Philosophy, Psychology and Education*. Wien: Passagen.

Oppenheimer, Todd. 2003. *The Flickering Mind. The False Promise of Technology in the Classroom and How Learning Can Be Saved*. New York: Random House.

Livingstone, Sonia. 2009. On the mediation of everything: ICA presidential address 2008. In *Journal of Communication*, 59(1): 1-18.

Lundby, Knut, ed. 2009. *Mediatization: Concept, Changes, Consequences*. Frankfurt a. M. u. a.: Lang.

McLuhan, Marshall. 1962. *The Gutenberg Galaxy: The Making of Typographic Man*. Toronto: University of Toronto Press.

Pachler, Norbert, Ben Bachmair, and John Cook. 2010. *Mobile learning. Structures, Agency, Practices*. New York u. a.: Springer.

Parmentier, Michael. 2009. Der Einbruch der Bildungsidee in die Sammlungsgeschichte. Auf der Suche nach den Ursprüngen des modernen Museums. In *Zeitschrift für Erziehungswissenschaft*, Jg. 12, H. 1:45-63,

Pastore, Erica. *The Future of Museums and Libraries: A Discussion Guide* (IMLS-2009-RES-02). Institute of Museum and Library Services. (2009) Washington, D.C. Retrieved July 19, 2010, from: http://www.imls.gov/pdf/DiscussionGuide.pdf.

Perger, Josef and Theo Hug. 2000. Transversale als "reine" Vernunft? – Ein Plädoyer für die Relativierung des Reinheitsgebots. Ethik und Sozialwissenschaften. *Streitforum für Erwägungskultur*, 11(1) : 132-134.

Roberts, Lisa C. 1997. *From Knowledge to Narrative: Educators and the Changing Museum*. Washington, DC: Smithsonian Institution Press.

Roth, Harriet, ed. 2000. *Der Anfang der Museumslehre in Deutschland: das Traktat "Inscriptiones, vel, Tituli Theatri Amplissimi" von Samuel Quiccheberg*. (Latin-German), Berlin: Akademie Verlag.

Rusch, Gebhard. Mediendynamik. 2007. Explorationen zur Theorie des Medienwandels. In *Navigationen. Zeitschrift für Medien- und Kulturwissenschaften*. Jg. 7, H. 1: 13-93.

Schäfer, Christoph. 2009. *Didaktik der Erinnerung. Bildung als kritische Vermittlung zwischen individuellem und kollektivem Gedächtnis*. Münster et al.: Waxmann.

Schmeer-Sturm, Marie-Luise. 1997 Musumspädagogik. In *Taschenbuch der Pädagogik*. Vol 3, edited by Helmwart Hierdeis and Theo Hug, 1092-1101. Baltmannsweiler: Schneider–Verlag Hohengehren.

Schmidt, Siegfried J. 2008. Media Philosophy – A Reasonable Programme? In *Philosophy of the Information Society*.

Proceedings of the 30th International Ludwig Wittgenstein Symposium, Kirchberg am Wechsel, Austria 2007, Vol. 2. Edited by Herbert Hrachovec, and Alois Pichler, 89-105. Frankfurt u. a.: Ontos.

Schmidt, Siegfried J. 2010. *Die Endgültigkeit der Vorläufigkeit.* Weilerswist: Velbrück Wissenschaft.

Schön, Donald A. 1991. *The Reflective Practitioner. How Professionals Think in Action.* Aldershot/Hants: Basic Books.

Simon, Nina. 2010. *The participatory museum.* Santa Cruz, Calif.: Museum.

Stogner, Maggie B. 2010. The Media-enhanced Museum Experience: Debating the use of Media Technology in Cultural Exhibitions. In *Curator: The Museum Journal*, vol. 52, no. 4: 385-397.

Turkle, Sherry and Seymour Papert. 1990. Epistemological Pluralism: Styles and Voices within the Computer Culture. In *Journal of Women in Culture and Society*, 16(1): 128-57.

Varis, Tapio. 2006. *eLearning and higher education.* Retrieved January 30, 2010, from: http://www.elearningeuropa.info/files/media/media11006.pdf

Vieregg, Hildegard. 1990. Meilensteine in der Entwicklung der Museumspädagogik. In *Museumspädagogik: Grundlagen und Praxisberichte*, edited by Schmeer-Sturm, Marie-Luise et al., 6-16. Baltmannsweiler: Schneider–Verlag Hohengehren.

Weber, Susanne and Susanne Maurer, eds. 2006. *Gouvernementalität und Erziehungswissenschaft. Wissen – Macht – Transformation.* Wiesbaden: VS Verlag für

Sozialwissenschaften.

Welsch, Wolfgang. 2000. *Vernunft. Die zeitgenössische Vernunftkritik und das Konzept der transversalen Vernunft.* Frankfurt a. M.: Suhrkamp.

Wetterlund, Kris and Sayre, Scott. 2009. *Art Museum Education Programs Survey.* Conducted from March through April 2009. Retrieved June 20, 2010, from: http://www.museum-ed.org/images/stories/research/2009medprogsurvey.pdf

The iPhone and Its Use in Museums

BJARKI VALTYSSON, NANNA HOLDGAARD
& RICH LING

IT University of Copenhagen

As with other cultural institutions, museums have always been ascribed specific roles in society. Throughout the years, these roles have not changed dramatically, even though the means to fulfill them have. As Tony Bennett convincingly documented in his book *The Birth of the Museum*, the museum was from its very beginning meant to enlighten and educate the public, while serving as the temple of the arts in an elitist fashion. Furthermore, the museum served as an instrument of the disciplinary society, where people from different social classes met. Ideally, the lower classes could learn from the "correct" behavior of the upper classes. Finally, in a similar Foucaultian fashion, the museum could filter out certain knowledge, thereby prioritizing certain historical readings, artifacts, and knowledge instead of other alternative options (Bennett 1995).

One could argue that in an overloaded information society, the field that the museum is supposed to cover has grown beyond control. However, the public museum has maintained its position as being precisely the filter that chooses what is to be preserved and communicated for future generations. The information society has not made that selection process easier; on the other hand, it also provides opportunities to reshape, reinterpret, and recommunicate museum collections to the public.

The advanced mobile phone, often referred to as a smart phone, offers interesting possibilities for museums, particularly in terms of communicating with museum audiences and encouraging different kinds of participation from visitors. As a gadget that crosses borders of different regulatory frameworks, and also combines different semiotic features – text, pictures, moving images, sound and gaming, GPS navigation, browsing, instant messaging, e-mailing, and last but clearly not least, talking – the smart phone pushes the notion of convergence culture (Jenkins 2006) to its limits. However, because of the converging characteristics of smart phones, the possibilities for museums' use of them for different purposes can also be considered as threats. This really depends on the context in which they are used and the role they are meant to fulfill.

The aim of this chapter is to take a further look at how advanced mobile phones, especially iPhones, can contribute to the goals of public museums, and to discuss the potentials and challenges for the use of mobile phones in a museum context.

In order to frame the discussion, we will apply a Scandinavian perspective[1] examining the political climate in which Scandinavian public museums operate, and accounting for data extracted from the traffic data of a larger Scandinavian telecom operator. While there are several mobile phones that fit the description of a smart phone, our set of data concentrates on iPhone users.[2] This corresponds with our understanding of museums' focus, as most of them channel their energies toward developing applications for the iPhone and not other types of smart phones. By comparing political wishes with the available apps and data traffic among iPhone users, we come a step closer to understanding whether museums should consider the iPhone as a threat, or as a tool with which to achieve their goals. Furthermore, even though the museum has many roles in modern society, we will be concentrating on those of dissemination and communication.

Scandinavian museums and the political climate

The political climate for Scandinavian public museums has increasingly moved towards involving visitors in defining and contributing to the museum's content production. This is a process closely related to notions of the experience society, where it is maintained that by engaging personally in a given cultural process, the actual experience (*Erlebnis*) is going to be greater and more meaningful (*Erfahrung*) (Boswijk, Thijssen, and Peelen 2007, 22 - 24). Digital communication, the internet, and more recently mobile communication and the use of smart phones, are considered to play an important role in such a user involvement. Buzzwords like *user-generated content*, *produsers* (a portmanteau of producers and users), and *creative audiences* are meant to indicate this shift from the individual as a traditional museum guest - as a passive observer. The goal has been to empower them into more active and participative users.

This focus on the user relates quite well to the role of public museums in Scandinavia. From this perspective, public museums are meant to collect, register, preserve, research, and disseminate (communicate) their collections and documentation. In this sense, we define the museum as a public museum in the Scandinavian tradition. When a further look is taken at the role of the public museum, the smart phone potentially contributes to the communicative context as the various apps are meant to enrich and enhance the visitors' experience, making the museum visit – on site or online – more memorable.

Official policy documents in Scandinavia encourage user involvement. They look positively upon the potential of digital and mobile communication, particularly in reaching quantifiable goals and reaching out to new segments of the public. From the viewpoint of cultural policy, this goes hand-in-hand with the wave of *quantification* that has occurred in recent years (Duelund 2003), where tools of New Public Management and results-based contracts are meant to guide museum directors in achieving their goals. These include attracting external sponsorship, increasing revenue, and raising the number of visitors. However, little attempt is made to explain and conceptualize what kind of use is preferable, why increased user involvement is positive, and in the museum context, what kind of use is considered appropriate.

Hence, seen from a policy perspective, there is an emphasis on attracting a new, trendy segment of the population. This *creative class* is often associated with the iPhone since this particular hardware has a certain cachet as well as the potential to act as a platform for outreach activities. Museums consider iPhone owners and users as a valuable public. By betting on the iPhone, museums respond to political wishes;

they introduce the *digital paradigm* and mobile communication into their realm and involve an attractive segment of the population. In theory, increased use of the iPhone in museums responds to many of their goals. One could argue that iPhone use caters to popular trends, that it interprets information and makes it accessible in a new, playful form, thus making the museum experience more meaningful and memorable. An iPhone app that, for example, provides illustrative information about the exhibitions will, in theory, attract more museum visitors. In using the iPhone, museums try to change their image. Another idea is that iPhone use makes museums attractive to potential donors; this, in turn gives museum directors a budgetary space within which they can operate, while using the iPhone's communicative potential to teach, inform, and stimulate discussions and comments.

The increased use of iPhones (and potentially other smart phones) therefore reflects the current political climate in Scandinavia. In practice, however, things are not looking quite so rosy. As our overview of available iPhone apps to be applied in museum context indicates, there are not many and most of them cannot be said to enhance users' experience in any significant way. What is even more alarming is the result of our data set, which indicates that iPhone users are neither all that many, nor do they use their smart phone as the museum apps intended.

Apart from these facts, which will be treated in more detail later, there seems to be a discourse in the Scandinavian museum milieu that indicates a certain commodity fetishism

(McGuigan 2009), in which the iPhone plays a large role. Smart, creative connotations are associated with the iPhone and its users, and museums expect these to rub off on their "dusty" image. Thus, critical discussions on the actual consequences of applying the iPhone are set aside for the political agenda – to attract more people to the museums. Discussions on the Disneyfication of museums are still very relevant. What about the museum's role as a filter? What about the expertise of curators? Will everything be reduced to the logic of a theme park: as long as it gives memorable experiences? What about the enlightening role of museums? To these questions, official Scandinavian cultural policy remains surprisingly silent, as the climate that public museums operate within is more focused on attracting new segments of the population. The question is whether the iPhone, its museum apps, and the users operating them provide the key to increased attendance figures.

iPhones in museums

With the emergence of digital technologies, the focus on multimedia tours on PDAs or mobile phones has risen in order to enhance the museum experience. However, handheld guides in museums are not a new digital invention. Educational materials have flourished in museum exhibitions for many decades, communicating and interpreting the intentions and meanings of the collections, exhibitions, and exhibited objects. With great inspiration from the American philosopher and psychologist, John Dewey, museums (starting with North American museums) slowly began to take an interest in education and produce educational material and activities. In 1952 the first handheld museum guide, an analog radio tour, was introduced at the Stedelijk Museum in Amsterdam, and the possibilities were widely discussed (Tallon 2008). Today, smart phones are considered the next generation of museum devices, as *the* platform for personalized museum guidance (assuming the visitors use their personal mobile phone in the museums). The development of the handheld museum guide has moved from paper pamphlets and brochures to the analog radio tour and the audio guide on *Walkman*, *Discman*, and MP3-player to the multimedia tours on PDAs, computers, and smart phones.

Visitor studies have shown multiple times that a museum visit is a social activity and event. Therefore, the main criticism of the handheld museum guide has been the antisocial behavior of the visitor when using a handheld guide with headphones or headsets: the visitor is in his or her own space with no or little room for social interaction, and any conversation is disrupted by the guide. Likewise, the guides are accused of making the visitors block the exhibition halls and galleries because all the visitors follow the same preselected tracks and tours. As a consequence, visitors become passive and inactive because the guides limit their interest to the content of the guide. In relation to the multimedia tours where the museum experience is remediated through a screen, the worst case scenario for the museum curators has been the distraction of attention from the exhibited objects to the screen.

A lot of the attention that the iPhone has attracted at museums is rooted in the fetishism related to the iPhone and the identity of the Apple brand. In general, the iPhone and many Apple products are the focus of users' wishes and desires. Users see the brand as chic, elegant, and up to date. In focus groups associated with ownership of the iPhone, a respondent noted: "iPhone's design is completely unique. It is stylish."[3] This fixation on Apple products is actively cultivated, and Apple has been able to nurture the image among several influential and highly visible groups.

Among museums, being the first museum to launch new ways of using digital technologies and media is very important. For many museums, it seems like *being first* is the primary rationale - part of a larger marketing and communication strategy, in which museums try to position themselves as the best museum in order to sell a specific image. When the Santa Barbara Natural History Museum released their iPhone application in 2009, executive director Dr. Karl Hutterer stated: "We are excited to be the first museum to release an app for the iPhone" (Lindberg 2009). In similar fashion the Van Gogh Museum declared in a press release: "Van Gogh Museum first museum on the Continent to launch iPhone application" (Van Gogh Museum 2009).

Another motivating factor for museums behind developing iPhone applications is outreach. With Apple's online iTunes store, museums can reach out to a number of users they would not be able to elsewhere.

Since the advent of the iPhone in 2007, there has been a buzz about the iPhone and its potential and use in the museum. The year after release, the iPhone was presented as the promising next generation device at the prominent *Museums and the Web* conference. In 2009, at the same conference, the iPhone in museum settings was almost a given fact and the conference hosted the how-to workshop *Programming the iPhone/iPod touch for your museum*. As of today, a number of museums have developed and experimented with the iPhone, and the expectations are still high.

A quick search in iTunes for "museum" gives 361 hits with iPhone apps and 32 iPad apps.[4] However, a large number of the apps are not related to or developed for actual museums (either public or commercial). In a closer look at the iPhone apps, only about 1 in 7 are developed by or for a public museum – none of which are from Scandinavian museums. Most museums have categorized their iPhone apps as education, which matches the traditional role of the museum as the filter that disseminates research and knowledge to the public. For example, the Musée du Louvre presents masterpieces from its large collection with details and information about the specific art piece, and the app functions as a zoomable digital catalog. Others have a more playful and entertaining approach (though still categorized as education). For example, the National Museum of Natural History's (part of the Smithsonian) first app, *MEanderthal*, transforms you (your photo) into a prehistoric you. The Van Gogh Museum in the Netherlands has developed an app called *Yours, Vincent: The*

Letters of Vincent Van Gogh. The application uses Van Gogh's personal letters to tell the story and thoughts behind the artist and the person.

Looking at the current landscape, museum apps are generally associated with larger museums (such as the Smithsonian, Musée de Louvre, Victoria and Albert Museum, Brooklyn Museum). At this point there are none to be found in Scandinavia, though several museums are developing iPhone applications. The available applications are often free to download. By way of comparison, the same search on "museum" at Android Market gives 55 hits. This includes all applications and also those unrelated to actual museums. Some museums like the Smithsonian Institution (the National Museum of Natural History) and Brooklyn Museum have made two app versions available, one version for the iPhone and one for Google Android.

The potential

Given this development, it is important to ask whether there is true potential for the development of iPhone apps. Are iPhone users more likely to consume information that is made available in this format? Is it worth developing these applications or are they simply a short-term fad that will soon pass? To answer these questions, we need to look at the behavior of iPhone (and more generally smart phone) users. This analysis was not done specifically in the context of museum use, but rather with general access to internet services via a mobile phone.

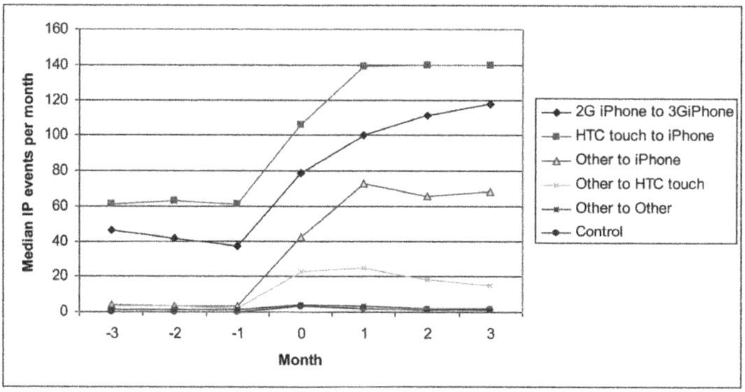

Figure 1: iPhone internet traffic

Given access to traffic data it is possible to test the idea that iPhone users consume more net-based services than other users. We had the advantage of access to a large number of iPhone users; since we could follow internet traffic consumption (in bytes) as the users changed to the iPhone 3G, we were able to see the effect of having a new iPhone. We were able to examine this transition using a quasi-experimental design.

The analysis took as its point of departure 33,400 individual users of the network of a larger Scandinavian operator. Apart from a control group of 4,500 individuals, to be included in the analysis an individual had to have changed handset during the trial period, which lasted for several months in 2009. Altogether there were approximately 20,000 iPhone users included in the sample. The vast majority of these (18,581) had switched from some other type of handset to an iPhone. In order to form the basis for comparison, other groups were included. Approximately 670 individuals had switched from an HTC Touch to an iPhone and 650 had switched from a 2G to a 3G iPhone. We included 7,297 individuals who had switched from any other (non-iPhone) type of phone to another phone as an additional control group.

As shown in Fig. 1 above, the users of iPhones indeed consume far more internet traffic than do other people. In the figure, the month that the individual made the transition to an iPhone 3G is noted as 0 in the Y axis. The months to the left of this are the period before the transition to the iPhone 3G and the months to the right are the period after the individual has made the transition to the new phone. The data shows clearly that the transition from other types of phones to an Apple 3G results in an increased use of mobile Internet.

Taking this finding a bit further, the owners of iPhones use them for all purposes – more than general users. They send more text messages and they often place more calls. Taking this increased use into account, they still are more net oriented than are general users. Approximately 90% of the communication events of a normal user are placing calls and sending texts. Only about 10% of the events are clicking on links. This is different for iPhone users. About 20% of the events are clicking on links. The remaining 80% are calling and texting.

These findings are good news for museums who wish to employ platforms such as the iPhone to perform outreach. The message of this analysis is that people who have iPhones (and to a much lesser degree other smart phones) use them to surf the web. This means that they will potentially use iPhones to examine content, including museum content.

It is possible to conceive the iPhone (as well as Android phones and other smart phones) as a platform for delivering information to museum visitors. The individual nature of the device means that it is possible to develop tailored presentations of material. The geo-location functionality can calibrate which part of the multimedia narration is relevant at a given time and the user can further indicate the simplicity or the sophistication of the presentation. Those with only a general interest can retrieve a simpler version of the material, while a person wanting special knowledge can access something that is more advanced. In the context of an art museum, for example, it is also possible that visitors might be interested in special "takes" on a particular era or set of paintings. They might want to know what music was popular at the time, the sophistication of painting equipment, the political situation, or the status of different populations. These could be woven into a more generic discussion of the objects in a particular location.

The two-way functionality of the mobile phone also allows users to add their commentaries and interpretations. It is possible to collect these in real time and perhaps display them for other museum visitors. This can be layered on top of other interpretations. There is also the possibility to "harvest" information: to determine which of the narratives are most popular and which objects elicit the greatest excitement among the public. How many people, for example, wanted to hear the history of the Mona Lisa versus the story about

Jackie Robinson's baseball mitt? It is also possible to think that such multimedia applications could serve as the basis for other tie-ins and the promotion of other exhibits or the generosity of sponsors.

Data on actual ownership and use

Studies of museum visitors have been completed in all the Scandinavian countries (Swedish Arts Council 2008; The Heritage Agency of Denmark 2010; Vaage 2009). According to these studies, the average Scandinavian museum visitor is female, over 50 years old, relatively well-educated, and most likely a school teacher. She prefers art museums, which she visits four or more times a year, and considers the museum visits as outings where she can spend time with her like-minded friends, though at times she also visits the museums alone.

The Heritage Agency of Denmark has carried out a national survey on Danish website users. The survey is the first of its kind in the Scandinavian countries and was launched due to the recommendation of a large museum communication report from 2006 by the Danish Ministry of Culture (2006). The characteristics of the average museum website user[5] are very similar to the average museum visitor but slightly younger. The average museum website user is female, between 30–49 years old, has a short or intermediate education, and lives in the capital area. She uses her computer at home to check the museum website for practical information such as opening hours and entry prices. In regard to social media and museums, she is very conservative and does not see the possibilities or understand how social media and museums can be connected (The Heritage Agency of Denmark 2010).

Unfortunately this profile is at odds with the standard user of an iPhone. According to the analysis of the iPhone owners shown above, they are significantly younger and more likely to be male when compared to general telephone users. Thus there is a mismatch between the traditional museum public and the users of iPhones.

Some of the museums are aware that iPhone users and museum visitors/museum website users are not exactly alike and have taken this into consideration in their development of applications. The Van Gogh Museum has specifically tried to adopt a non-traditional museum tone and has focused on creating an exciting experience to hold users' attention in their application *Yours, Vincent: the Letters of Vincent Van Gogh* (Peereboom, Schreurs & de Vet 2010).

Possibilities and barriers

There are possibilities associated with the iPhone and other smart phones when considering their use in the outreach and interpretation activities of museums. To be sure, they provide an interesting and a relatively easy platform with which to address the museum visitor and user. The smart phone allows the museum to develop digital material that is accessible to users. The material is accessible to users and can be accessed when and where users wish and can even be linked to the specific location of an exhibit or an artifact. iPhone applications can be tailored to the needs and interests of the individual and can go into depth or remain superficial just as the user wishes. Furthermore, they can access a variety of peripheral topics. In addition, this technology allows the museum to gather the insights and comments of the users. It can be linked to other forums and, for example, social networking sites. In this way, the technology can help facilitate engagement with the material of the museum, which can help the museum to fulfill its mission.

The next question is the degree to which the iPhone has been adopted broadly in society. As noted above, there are some groups that are more eager in their acceptance. However, just how eager is society when it comes to the adoption of the iPhone? The bad news is that there are actually quite few such devices, there is a skew in their use, only a small number of apps are available, and the museum public does not necessarily match the users of advanced mobile phones.

There are actually quite few iPhones. In the example above, only about 5% of the people in the whole network had an iPhone. This is in a portion of the world that has the highest income *per capita* and is very advanced in terms of mobile communication. Looking worldwide, only about 1% of phones are iPhones. Thus, while this type of phone is slowly being adopted (indeed there are many advanced phones being sold), there is a huge number of older devices still in use. Many people become comfortable with a particular mobile phone and are reluctant to give it up. In the network described above, the most popular mobile phone until quite recently was an eight-year-old out of production Nokia model. It was simple to use and many people, particularly older users, had become comfortable with it: they had the contacts they needed and it served their purposes. In addition, they perhaps no longer felt the need to assert their status with this type of artifact. This reluctance to adopt the newest technology means that these people will not be able to use any mobile internet applications that might be developed for use in museums.

It is also true that only a small number of apps ever get serious use. Most users have many apps downloaded onto their phones, but they use just a small number. There are thousands of apps available for the different platforms. It is difficult for an app producer or sponsor to get noticed in this jungle of possibilities.

Smart phones have a limited reach and, ironically, they can facilitate social connections that are at cross-purposes to those of the museum. The first of these issues has been discussed above. Looking for the moment at iPhone users, they are more active consumers of mobile internet than others. The material described here is clear on this point. This said, the consumption is not evenly distributed. An exceedingly small group consume the preponderance of traffic. In addition, these users tend to be younger and more often male. This is not the core museum audience. Furthermore, there are not that many iPhone (or more generally smart phone) users in the population. It is true that smart phones are very popular and often come quite high in sales statistics. However, there is a large reservoir of more traditional phones still in active use. These are being replaced with phones that are often more advanced, but the new phones are not always smart phones.

There is another dimension to the development of digital mobile-based material for museums. Specifically, there is a double-edged issue associated with user-generated content. There is often the perception that curators have special knowledge of the objects in a museum; the curator will be able to judge which objects are to be collected to represent a culture, movement, or era. There is also the perception that they will be able to document the context of a particular piece or trend.

Giving voice to "novice" museum content signals an openness and a willingness to accept alternative interpretations of cultural heritage. At the same time, it questions who has the correct interpretation of the cultural heritage. Museums often create displays that seek to gain the public's attention by taking up themes that provoke thought. At the same time, mobile telephony allows collections of interested individuals to quickly organize themselves (Rheingold 2002). If the provocation by the museum is too much for some groups, they can organize protests against the display, or groups could organize alternative interpretations of the display. Looking at the former, had the *piss Christ* controversy happened with the current technology on the scene, those whose sensibilities were abused by the photo could have used the internet and mobile communication to protest the event (Pertierra et al. 2002; Rafael 2003; Rheingold 2002). Mobile technology opens up the possibility for different groups to develop and spread their interpretations of cultural or scientific material. Museum visitors can download and deploy material while they are actually in the museum, thus testing the expertise of the institution's curators. The potential of digital communication allows museumgoers to organize alternative interpretations of museum objects: in a museum describing evolution, believers in creationism could develop their own narrations that contradict the line of thought suggested by the museum's own authorities.

Indeed, this might be the reason why a majority of the museum apps do not really encourage the user-generated content of the *creative audiences*, but rather are ways of branding the museum and disseminating (communicating) museums' digitized content. By doing this, museums are catering to political trends without losing their control as "gatekeepers."

Conclusion

iPhone users are an advanced user group that has a social cachet. They use a lot of data and there is a trendiness associated with apps for the iPhone. The threshold for developing this type of outreach is relatively low and because of this there is the ability to capture the interest of new groups. In addition the use of these services potentially allows for new types of participation and data collection.

This said, the eventual development of apps will be for a limited audience. There are hundreds and thousands of other apps competing for the user's attention. In addition there is a mismatch between the users of iPhones and the traditional museum visitor. Looking broadly at iPhone users, they are very active in their use of data services but despite the hype associated with smart phones, there are not many users overall.

The move to mobile material is also a double-edged sword. Use of mobile information increases the opportunity for outreach, but also means that museums have, to some degree, given up control of the message. The gray zone that creates provocation has been reduced. Digital technology gives those who disagree a new way of making their position known.

In sum, advanced mobile phones can contribute to the goals of the museums. It is worth trying out new approaches since it indicates to the public that the museum is aware of these trends and interested in engaging with and involving the public. There are significant possibilities as well as significant barriers associated with these developments. Not many people have iPhones and museum visitors are not often iPhone users. This said, it can be a way to pursue the goals of the institution if done with a realistic sense of the possibilities and the barriers.

Notes

1. Scandinavia is the northern region in Europe and includes the following countries: Sweden, Norway, and Denmark.

2. In this paper we concentrate on the iPhone though it is not the only advanced phone that could be used. Other smart phones such as Android, Blackberry, and iMode phones have these capabilities.

3. These focus groups were held in 2009 in Norway. They were a series of two groups among people who owned iPhones. Approximately 8 persons were included in each of the sessions. The material was recorded, reviewed for relevant categories, and sections of the material were transcribed.

4. The numbers are extracted from iTunes in mid-August 2010.

5. The data on the museum website users is Danish, however we expect the Norwegian and Swedish museum website users are much like the Danish due to the big similarities between the Scandinavian countries and the museum visitor characteristics.

References

Bennett, T. 1995. *The birth of the museum: History, theory, politics.* Routledge.

Boswijk, A., T. Thijssen, and E. Peelen. 2007. *The Experience Economy: A New Perspective.* Amsterdam: Pearson Prentice Hall.

Danish Ministry of Culture. 2006. *Udredning om museernes formidling.* Danish Ministry of Culture, Copenhagen.

Duelund, P. 2003. *The Nordic Cultural Model: Nordic Cultural Policy in Transition.* Copenhagen: Nordic Cultural Institute.

Jenkins, H. 2006. *Convergence Culture: Where Old and New Media Collide.* New York: New York University Press.

Lindberg, E. 2009. Natural history museum launches iPhone app. In: *Daily Sound* June 19, 2009. Consulted 1 September 2010 http://www.thedailysound.com/061909museumapp

McGuigan, J. 2009. *Cool Capitalism.* New York: Pluto Press.

Peereboom, M. et al. 2010. Van Gogh's Letters: Or How to Make the Results of 15 Years of Research Widely Accessible for Various Audiences and How to Involve Them. In J. Trant and D. Bearman (eds). *Museums and the Web 2010: Proceedings.* Toronto: Archives & Museum Informatics. Published March 31, 2010. Consulted 1 September 2010 http://www.archimuse.com/mw2010/papers/peereboom/peereboom.html

Pertierra, R., et al. 2002. *Txt-ing selves: Cellphones and Philippine modernity.* Manila: De La Salle University Press.

Rafael, V. L. 2003. *The cell phone and the crowd: Messianic politics in the contemporary Philippines.* Public Culture 15:399-425.

Rheingold, Howard. 2002. *Smart Mobs.* Cambridge, MA: Persius.

Swedish Arts Council. 2008. *Museer & konsthallar 2008*. Swedish Arts Council, Stockholm.

Tallon, L. 2008. Introduction: Mobile, Digital, and Personal. In Digital Technologies and the Museum Experience: *Handheld Guides and Other Media*, ed. L. Tallon and K. Walker. Lanham, MD: Altamira Press.

The Heritage Agency of Denmark. 2010. *Netbrugerundersøgelse*. Kulturarvsstyrelsen, København.

Vaage, O. 2009. *Mediabruks undersøkelse, 2008*. Statistics Norway, Oslo.

Van Gogh Museum. 2009. *Van Gogh Museum first museum on the Continent to launch iPhone application*. Consultet 1 September 2010 http://www.vangoghmuseum.nl/vgm/index.jsp?page=201729&lang=en

Web Empowerment for Cultural Heritage Organizations: Towards a Theoretical Framework

MONICA MURERO

E-Life International Institute

University of Florence

What does it mean for cultural heritage organizations (CHOs) to embrace social media? What criteria, strategies, and resources should move CHOs to play an active role in interacting with the public via Facebook, Twitter, or vlogs posted a few minutes after visiting the museum? More importantly, when should CHOs embrace Information and Communication Technologies (ICT) innovations, or avoid them? What criteria and methods should be followed to evaluate risks and opportunities in adopting Web 2.0 innovations for the museum?

Nowadays, it is mundane to suggest that CHOs, both public and private, are intrigued by innovative opportunities offered by the diffusion of ICT, and in particular social media. Recently, Web 2.0 has offered new services and multimedia applications to millions of users around the world. Social media is challenging the way people interact inside and outside cyberspace. In other chapters the term social media has been extensively defined and analyzed (see for example, Theo Hugg's chapter). We should recognize that at the time this book is written, social media is becoming new media history, being one of the most emerging realities in both business and not-for-profit organizations. Web 2.0 opportunities and possibilities are attracting marketing study, media analysis, and interdisciplinary research in all sectors of society. (O'Really, 2005; Hansen 2006; Jenkins, 2006; Murero, 2010).

We are at the infancy of the introduction of innovative solutions for museums that will completely change the classic concept of the audio guide in cultural education, thanks to new ICT technologies, services, and applications that facilitate human interaction. The case for embracing or avoiding social media or ICT-based innovations has both advantages and caveats. For example, one advantage is the fact that several studies have shown that interaction and communication processes among networks of visitors have a positive impact on the educational and learning experience. Falk and Dierking have highlighted that the cultural experience is influenced by the social context that includes interactions among visitors. In the field of CHOs, a limited number of innovators are using social media such as wikis, micro-blogs, and social networks to interact with online participants before, during and after the visit. Social media enhances the online educational experience and interactions among thousands of fans, and contributors are creating unprecedented challenges for organizers and curators of cultural exhibitions, which might affect organizers' decision making processes. For example, the curators of Rome's National Museum of the XXI Art, Maxxi, which opened in Spring 2010, decided to embrace online interactivity while the building was still under construction; they received hundreds of inputs every day from online fans who exchanged opinions, gave suggestions, and participated in a collective effort for the realization of the new museum. Online visitors were able to preview some objects in the collection, and perform some interactive activities, like ludic experiences consisting of recomposing a puzzle of a particular artistic object piece by piece. People were asked to recompose the "real" object that was not yet exposed to the public and visible only online. Visits to the museum's website generated thousands of contacts during the months before the opening of the exhibition. Several thousand fans followed the day-by-day advances of the new museum online, before it was inaugurated. Today, after opening to the public, the Maxxi's page on Facebook alone counts about 12,000 fans commenting and exchanging information about the museum's events.

We are still at the beginning of a phenomenon that is expected to grow in the future. Social media services and features could offer unique opportunities for users to enhance, share, and better memorize the learning experience, particularly when associated with the use of mobile devices and multimedia experiences (Safko and Brake, 2009; Murero, 2010). Concrete benefits are reported before, during, and after the visit to the museum. Some authors maintain that social media adds value to cultural institutions, also in epistemological terms (Russo et al., 2006).

The first question we address in this chapter is: are cultural heritage organizations successfully exploiting the opportunities of social media and mobile devices to deliver a high quality service to users?

Nowadays, social media matter. The solutions that the cultural heritage sector has developed so far, many of which are described in this book's pages, are interesting not only to individual institutions but also to the cultural heritage sector as a whole. Successful applications might influence and shape the future ways in which cultural institutions act. Early adopter solutions today might become a source of inspiration for many complex cultural interactive experiences tomorrow.

Another crucial question that this chapter addresses is the following: are CHOs considering the potential risks associated with the use of social media? The case for embracing or avoiding social media or ICT-based innovations has caveats as well. One caveat, for example, is the fact that several studies have shown that spontaneous groups – virtually anyone with internet access – can easily generate online content in different forms, which are not always welcomed by cultural institutions. As a result, the products of social media generated within the context of cultural events are easily available online, and could circulate among millions of users without any control from the hosting cultural organization. This is one of the most threatening phenomena that discourage organizations from adopting social media. There are several other risks in adopting ICT innovations and social media, which pertain to the sphere of internal organization, particularly in terms of lack of skills and insufficient allocation of resources. Online content ownership and legal implications should be taken into consideration as well, when using social media to interact with the public. There is a need to professionally evaluate whether to start using ICT-based innovations.

Cultural heritage organizations already embracing Web 2.0 devices and applications are mainly located in North America and northern Europe. Some use viral marketing techniques to increase the number of visitors; others offer innovative solutions enhancing the learning experience – as we have observed in this book.

In southern Europe, one of the richest areas of the world in terms of cultural heritage, the majority of CHOs are struggling every day with insufficient funds to preserve and conserve unique masterpieces. CHOs could largely benefit from Web 2.0 and ICT innovations to enhance the educational experience; however, they are among the weakest adopters. Although it is estimated that 80% of the entire worldwide cultural heritage treasure resides in Italy, the majority of CHOs face several difficulties in adopting innovations. In the light of successful experiments, like the Maxxi museum in Rome and other best practices presented in this book, cultural institutions could largely benefit from the adoption of social media and innovative ICT devices, services, and applications, when favorable conditions apply.

Are CHOs using a professional approach and methodology to evaluate if the organization is ready for Web 2.0 challenges? How can CHOs treasure best practices, learn from previous mistakes, and find valid guidance for practitioners and internal staff who are taking the first steps into Web 2.0 innovation?

CHOs need practical criteria and methods to evaluate the risks and opportunities in adopting 2.0 innovations

Although many cultural organizations around the world are interested in adopting and experimenting with social media and ICT-based innovations, one of the problems for CHOs is translating intentions into actions. Empirical studies (Murero, 2010, ongoing) show that a lack of knowledge about new technologies among cultural heritage organizations is a very widespread reality. Cultural organizations often suffer from limited ICT and new media literacy, poor knowledge of communication processes, little or no use of methods to assess risks and opportunities when evaluating innovation, lack of specialized human resources, limited funds, and in some cases, internal/external barriers and resistance. Long restrained by limited resources and services, cultural heritage institutions and policy makers, particularly in the public sector, could be challenged if not overwhelmed by innovations that might translate into unmanageable situations for a Web 2.0 unprepared CHOs.

In the majority of cases, cultural heritage organizations need to move from the excitement of dealing with a great opportunity, and transform it into a concrete reality. Cultural heritage organizations need a framework to help them assess risks and opportunities and make an informed decision about whether or not to embrace the social web. In this chapter we'll propose a pragmatic approach that offers methodological and practical solutions to these current issues.

We'll focus on social media for practical reasons, while we keep in mind that the framework can also be applied to ICT-based innovations like mobile devices, and to other sectors besides the cultural heritage field.

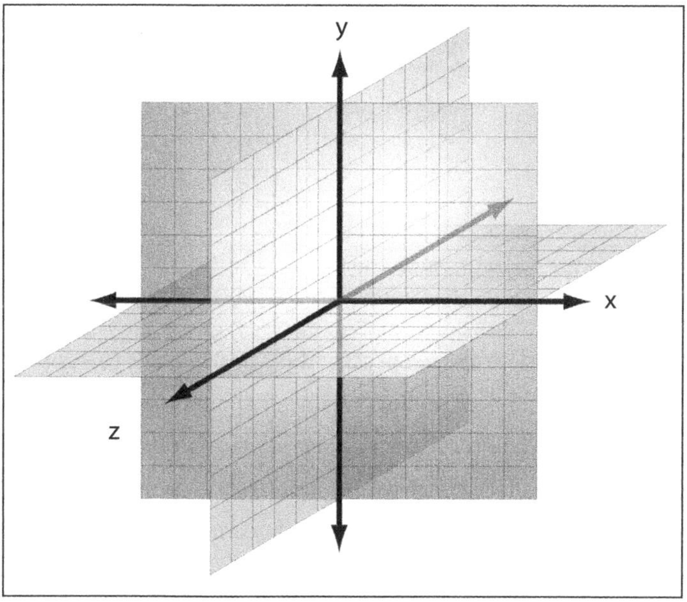

Figure 1: A Cartesian representation of the CHO2.1 Framework, where x=Objects y=Agents z=Endogenous-Exogenous Variables.[1]

Towards a framework for the cultural heritage

In this section we'll focus on a new framework for the cultural heritage sector, the Cultural Heritage Framework 2.1 (CHO2.1).

The Cultural Heritage 2.1 (CHO2.1) is a tridimensional and dynamic longitudinal framework for assessing risks and opportunities in adopting ICT-based innovations, including social media, mobile devices for enhancing the learning experience, and other ICT-based innovations for museums.

It offers a method to evaluate whether the introduction of innovation is sustainable for the organization in terms of capabilities, resources, and processes; once this works,

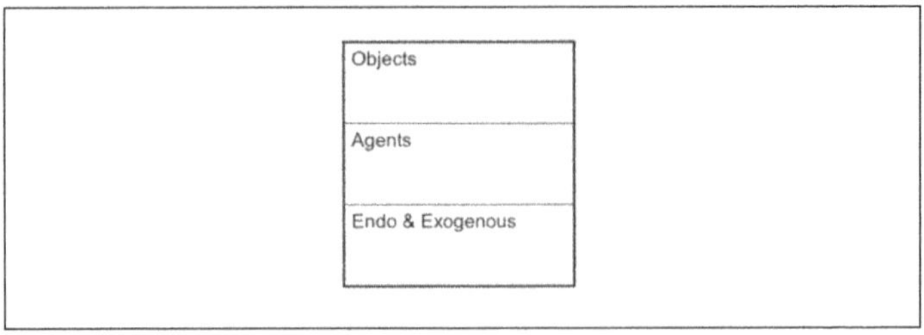

Figure 2: The Tridimensional Framework CHO2.1

the framework proposes practical planning tools to support and guide decision makers through the adoption process, particularly when cultural organizations rely on (scarce) public funds. Based on ongoing empirical research (Murero, 2010), the framework should be updated over time.

Why 2.1? The number 2.1 refers to the release number. In informatics terms, a release number is used to differentiate one software version from the precedent, available to its users. In this case, the framework 2.1 refers to Web 2.0 and intends to become the "next release" available to cultural heritage organizations.

CHO2.1 consists of three main dimensions: Objects, Agents, and Endogenous and Exogenous variables, as we'll describe. (Figs. 1 and 2)

The first element of the framework is *Objects*. It refers to goals, ICT innovations that the CHO intends to adopt, targets, information, action plan, and more. The dimension Objects includes several subcategories. It also offers a methodological approach to assess the risks and opportunities in adopting ICT-based innovations, by considering all available resources and barriers. The second dimension of the framework takes into consideration all the *Agents* – the human resources involved in the management and decision making – and assigns responsibilities to develop the project. It also considers the parties that act outside the control of the museum, but might influence its decisions: such as online users using social media to create blogs or fan pages within social networks that might impact CHO activity.

Examples of Agents are: internal resources or team, third parties, CHO partners and consultants, policy makers, and visitors interacting within social networks or other social media. Finally, the third dimension considers *Endogenous and Exogenous* variables that might impact on the CHO activity. Examples include internal obstacles and cultural resistance, administrative and legal barriers, media pressure, and other "unforeseen" events.

How the CHO2.1 framework works

The three dimensions of CHO2.1 – Objects, Agents, and Endo-Exogenous variables – dynamically interact and influence each other. For example, if the museum plans to introduce video-blogging (vlog) at the end of the visit to the museum, the CHO2.1 framework will help decision makers evaluate the feasibility of all the operations involved (objects: goals, targets, technical issues, strategy, incomes, costs, legal and administrative barriers, etc.). The Agents dimension will dynamically help focus on human resources available for each *object*, define their cost and responsibilities, and if it is wise to hire new resources or rely on third parties. For example,

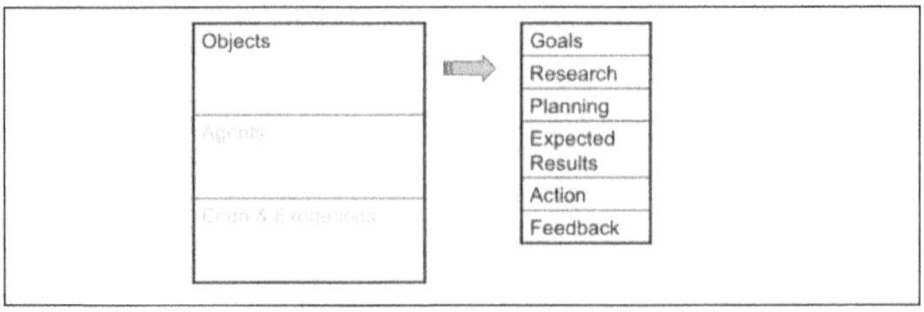

Figure 3: Objects

the framework will report on the third dimension if there are administrative or legal impediments or barriers coming from Endo-Exo variables. Once all dimensions are fully reported, the CHO2.1 framework will offer to decision makers a valid instrument to evaluate the ROI (return on investment) of the innovative ICT solution and whether adoption is wise. In some cases, the fulfilment of the framework's dimensions and detailed subcategories might suggest that there are not enough resources available to adopt a certain solution. Even in this case, all information collected in the CHO2.1 framework can be used to update the project at a more convenient time. During the CHO internal staff meeting, one person should take notes and report the decisions in electronic format, for later evaluation.

As we have mentioned above, the three dimensions and subcategories of the CHO2.1 Framework influence each other. With this in mind, we'll start with the Objects dimension and its subcategories; later we'll analyze the other two dimensions of the framework. We'll examine each one of these three dimensions and their subcategories in detail, and learn how to use the framework in practice.

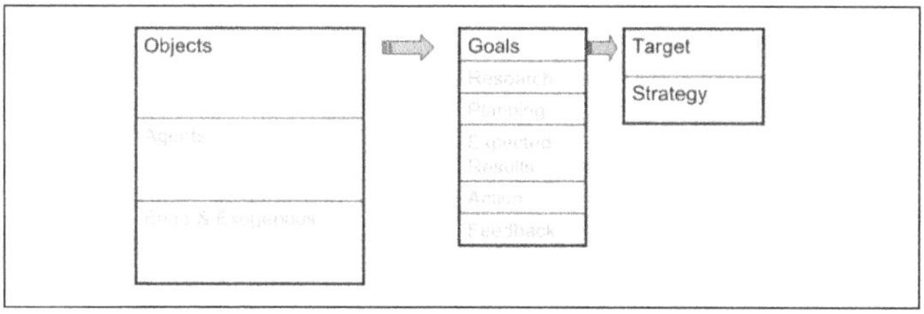

Figure 3: Goals

FIRST DIMENSION: OBJECTS

The first dimension of the CHO2.1 framework is Objects. (Fig. 4) It includes five subcategories or phases: Goals, Research, Planning, Expected Results, Actions and Feedback. These subcategories help museum staff organize activities and implement ICT innovations if conditions allow for it, by following a methodological approach, from goals to feedback. For example, if the CHO decides to open a page on Facebook with the goal of increasing the number of young visitors per year, CHO will report in the "object" dimension of the framework: this goal, all information related to the target, the chosen strategy, the best practices others have already adopted, the resources available - also in terms of incomes and expenditures to finance the operations - and the steps that need to be taken to translate goals into an action plan whose results will be evaluated during the feedback phase, and eventually corrected. In the next paragraphs we'll see the Objects' five subcategories.

Objects – goals: target and strategy

One of the first questions CHOs should ask when approaching

ICT-based innovation is: what do we intend to achieve? A summary of the intended goals should be documented here. (Fig. 4) Examples of goals might include (but are not limited to): increase the number of visitors by adopting social media, increase awareness and brand image of the CHO, and increase ROI (return on investment) after introducing an innovative ICT-based solution that visitors can access via their smart phones.

Another question that goes along with the definition of the goals, that is also very important if not crucial for the entire planning process, regards the target. The public of the cultural organization is very important. CHOs should take into consideration the museum's statistics on visitors in the past three years, as well as prospective visitors. The definition of the target should be in line with the definition of the goals. It is very important for the success of operations that the goals and target are coherent. For example, if the cultural heritage organization attracts only tourists every year, the staff should aim at this target with dedicated actions. If the CHO needs to evaluate if it is feasible to attract local schools, research should be done and separate actions should be planned and customized – once the reasons are clarified why locals do not attend the museum. The "strategy" is the third element that should never be separated when discussion of goals and targets arises. A summary of the strategy – also in terms of "media strategy" – to address the target, should be documented here. Social media must be a coherent solution to strategically reach the desired target, within a coherent

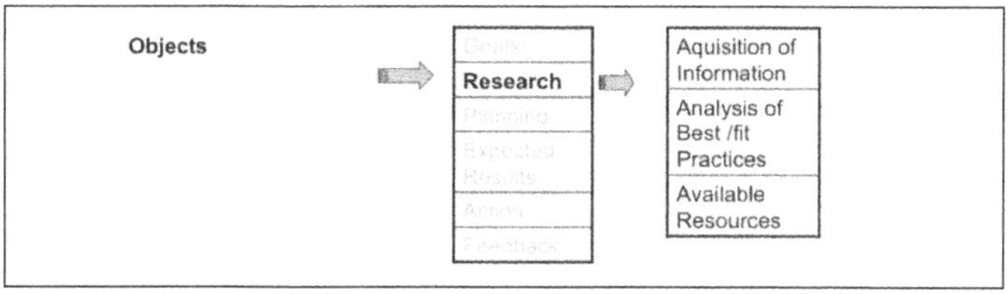

Figure 5: Research

mix of choices that we'll specify later. For example, if the museum is located in a remote area where the penetration of ICT is very low and the internet is not diffused among the intended target of local visitors and schools, it would not be advisable to invest in social media for that particular public – whereas other actions and media plans would be more successful in reaching that target. These examples introduce us to the second subcategory of the Objects dimension that should be taken into consideration to define goals, targets, and strategy: Research.

Research: information, best practices, and available resources

During the Research phase (Fig.5), the CHO gathers all useful information to evaluate risks and opportunities related to the adoption of the innovation. For example, the CHO explores the characteristics of the innovation, the available alternatives, and the impact that the adoption might have on the organization processes, including costs. The ultimate goal of the Research phase is to evaluate if favorable conditions apply so as to experiment and eventually adopt the innovation – in consideration of the unique characteristics of the CHO in terms of resources, internal processes, skills, and limits.

Subcategories include: Acquisition of Information, Analysis of Best Practices, and Available Resources.

Acquisition of information: The results of the research and information should be collected, shared, and reported here. Sources of information might include internal and external professional opinions – see the Agents dimension in the next paragraphs. The analysis of previous studies, including unsuccessful experiences, should help clarify the characteristics of the innovation and the implications of its adoption for the organization.

Analysis of Best Practices (or Practices that Fit): When using this framework, excellent practices should be recognized, documented, and implemented when possible. The CHO should carefully consider the solutions that apply to its reality. In fact, *Practices that Fit* refers to the best experiences and solutions in consideration of the CHO's characteristics, resources, and barriers. Such solutions might fit the CHO reality better than other best practices, and should be carefully identified.

Available Resources: All available internal and external resources should be documented here, including incomes, cultural expertise, technical appliances and support, external parties, and funds. As we have previously observed, any subcategory of the framework can benefit from the inputs of other dimensions and subcategories. For example, in this case, the subcategory Available Resource crosses the Actor (human resources) dimension when considering the

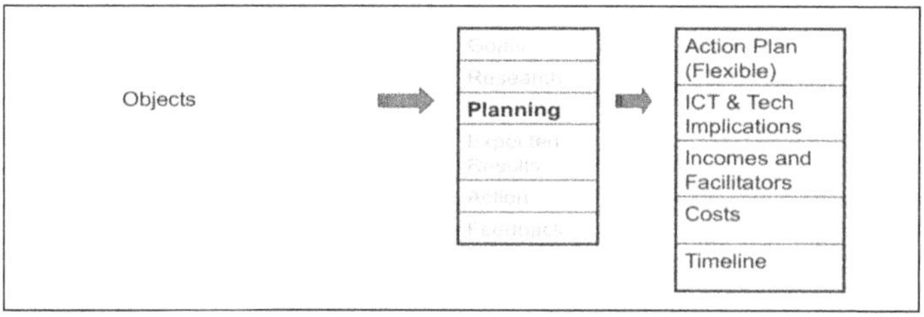

Figure 6: Planning

necessary skills to adopt the innovation. Results should be reported here.

The Research phase is very important for the evaluation of the processes involved with the eventual adoption of a new technological solution or object. At the end of the Research phase, crucial decisions should be made. Once all useful information is recorded, the framework should offer to the CHO the possibility to review and when necessary change the initial goals, target, hypothesis, and strategies. However, if results show that the conditions to proceed are not yet mature, then the CHO should postpone the project and only reconsider it when conditions improve. If results show that favorable conditions apply, the CHO can proceed with the next phase of the Objects CHO2.1 framework: Planning. (Fig.6)

Planning: action, implications, costs and incomes, timeline

With the successful conclusion of the evaluation phase, the CHO could decide to implement the innovation. In this case, the CHO should now have all the elements to develop the next phase of the adoption process: Planning.

During the planning phase, the CHO translates goals and

strategies into practical actions, prepares the budget, and sets a timeline. The budget must include all incomes and costs of the operations, and must be kept constantly updated. All available resources should be allocated in the budget by considering the decisions made at the end of the Research phase, and possible different scenarios. The Planning phase includes five subcategories: Action Plan, ICT implications, Incomes and Facilitators, Costs, and Timeline.

Action Plan (flexible): When using the CHO2.1 framework, a list of actions to implement the Social Web (or other ICT innovations) in the organization should be documented here. The Action Plan should be flexible enough to welcome changes over time, on the basis of emerging needs and new scenarios occurring in any of the other dimensions and subcategories of the framework. The Action Plan should include a media and communication plan, particularly important when adopting an ICT-based innovation.

ICT (implications): Technological matters are often unclear to museum managers, according to ongoing research (Murero, cited above). Technical needs and implications involved in adopting an ICT-based innovation – including, for example, interoperability, but also data ownership issues related to the contents exchanged via social media – should be carefully documented here.

Incomes and Facilitators: Incomes and Costs – the former described in the next paragraph – should be precisely included in the Budget. The two subcategories of incomes and costs are so important that they are

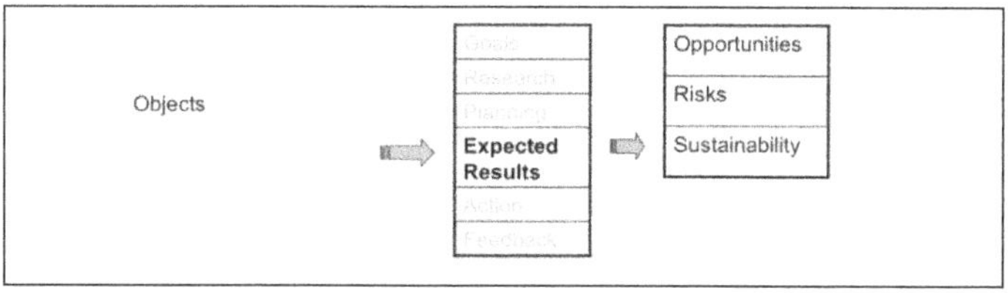

Figure 7: Expected results

exclusively identified in two well-identified objects of the framework. When using this framework, all incomes should be integrated in the budget, including facilitators or incomes from donations, sponsorships, and other revenues.

Costs: Foreseen costs of each element in the budget of adopting the innovation should be documented here. An extra 5% calculated on the basis of the total costs should be planned for "unforeseen".

Timeline: The plan's timeline should be set to keep operations under control and proceed as scheduled during the Action plan – as we'll see in the next paragraphs.

As per every subcategory of the tridimensional framework, it is important to keep in mind that further major inputs could dynamically impact the Planning phase. For example, the second dimension of the framework, Agents, will clarify who does what, in terms of human resources and responsibilities for each planned action, and should be documented here. Moreover, Endogenous and Exogenous variables could limit or facilitate single elements of the plan – as we'll observe in detail later – and should be reported here. In the next paragraph we'll focus on another phase of the Objects dimension: Expected Results. (Fig.7)

Expected results: opportunities, risks, and sustainability
Here, cultural organizations should define and report what they expected to achieve in practical and clear terms. Strictly related to the planning phase, the definition of the expected results helps set benchmarking criteria – useful for a later evaluation of results – and keep the operations under control. The Expected Results phase includes the three important subcategories: Opportunities, Risks, Limitations and Legal Implication, and Sustainability.

Opportunities: Perceived and measurable positive results from using social media or an innovative ICT solution should be reported here. Examples of Expected Results include: 10% more visitors per year thanks to word-of-mouth generated via Facebook page.

Risks, Limitations, and Legal Implications: A summary of the expected threats and barriers involved with the use of ICT-based innovative solutions, including foreseen negative scenarios and legal implications, should go here. Corrective measures should be planned here as well, to overcome negative outcomes or events.

Sustainability: Managers of the CHO should realistically consider the innovation's capacity to endure over time, due to the planned actions, the innovation's characteristics, and available resources of the organization. Expectations in terms of sustainability should be documented here.

A practical example of how to report expected results in the framework could be the following: Continuing with our previous example about the creation of an official Facebook

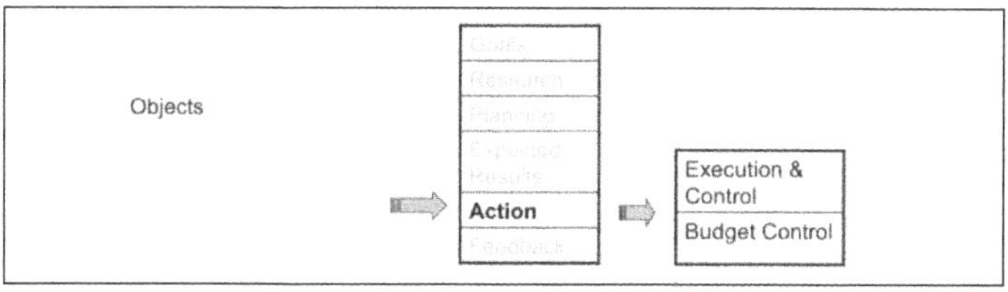

Figure 8: Execution and control

page, the managers of the CHO could report that they expect 10% more young visitors after the first year of online activity via the social network, and 12% more the second year. Regarding sustainability, for example, the CHO will report that thanks to a special two-year grant received from an external sponsor, they expect to hire dedicated staff to interact with online visitors. However, the contract will be for a period of two years. Expected incomes generated from extra visitors per year will finance the cost of the staff for another two-year period.

Action: execution and budget control

The Action phase consists in the practical realization of all the activities planned according to the goals, target, strategies, and timeline. The Action phase can be performed only if all the previous dimensions and subcategories of the CHO2.1 Framework are fully considered and reported. For now, and for explanatory purposes, we'll proceed with the analysis of the Objects dimension, and later we'll analyze the other two dimensions of the framework. With this in mind, the Action phase includes two subcategories: Execution and Budget Control. (Fig.8)

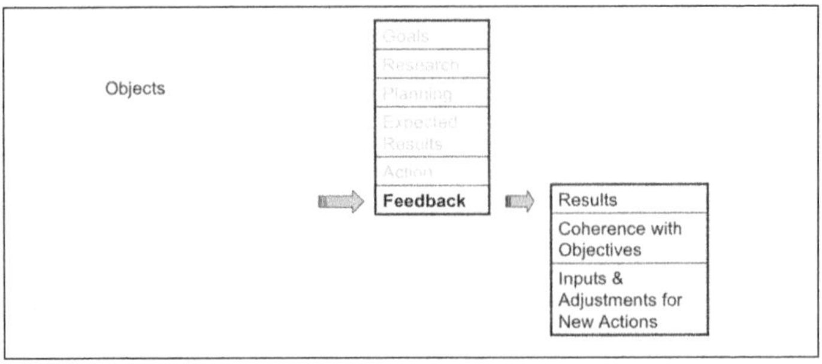

Figure 9: Feedback

Execution and Control: Actions should be carried out using allocated resources according to the actions planned, budget, and timelines, and supervised accordingly. Every deviation from the scheduled plan should be promptly reported and corrected.

Budget Control: During the "Action" phase, the budget's expenses and incomes should be done under strict control, and corrective actions must be introduced, when necessary. Ad hoc Agents should conduct planned Actions and supervision of the "Action" phase.

Feedback: analysis of the results, coherence, and adjustments

The Feedback phase (Fig.9) consists of the analysis of the operations undertaken during the Action phase. The Feedback analysis includes an evaluation of the adherence of results to the plan, with the intended goals and expectations. This is an important learning phase for the CHO that should critically analyze results, best practices, and mistakes. If mistakes are

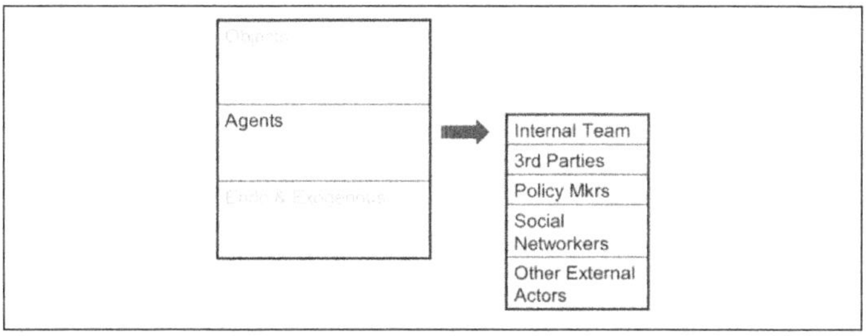

Figure 10: Agents

observed, they should be corrected as quickly as possible. The results of the Feedback phase should update the CHO2.1 framework and introduce future actions and goals.

SECOND DIMENSION: AGENTS

The second dimension of the CHO2.1 framework is called Agents (Fig.10) and takes into consideration all human resources involved in the process of evaluation and adoption of an ICT-based innovation. The Agents dimension helps management highlight its needs in terms of human resources, skills, and services, and planning for future activities. For example, the CHO should report here if there is a need to hire internal resources with specific skills, or exteriorize services and activities to third parties. If the CHO decides to take an active role in social media, for example, then it should allocate adequate resources to interact with the public online.

The CHO should report all possible barriers coming from human resources in adopting the innovation – including, for example, internal staff resistance to change – and, when possible, corrective actions. The Agents dimension consists of two main categories, internal and external parties, which are specified in five subcategories: Internal Team, Third Parties, Policy Makers, Social Networkers and Other External Actors.

Each subcategory presents specific characteristics and offers different opportunities and challenges in terms of human resources and skills, but also presents limits: resistance to change and risks to the organization's management are only a few examples. We recognize that several actors intervene in the framework with different levels of responsibilities and activity. The range of individual and group attribution of power should be documented here. The framework will offer a clear identification of the number and typologies of resources necessary to address the project's needs, and clarify who does what with different levels of responsibilities, starting with the *Internal Team*. The Actor's dimension must be updated or integrated as necessary, over time.

The Cultural Heritage organization cannot directly control external parties. However, Third Parties – partners and consultants, for example – or Other External Agents can offer services and collaborations that the organization cannot afford, does not have, or might not be able to perform internally. For example, reliable third parties or other external agents can help the CHO take correct investment decisions on ICT matters, particularly when innovations involve high costs, special equipment, and skills that the CHO does not have internally. The CHO does not control policy makers, the general public acting as Social Networkers, and Other

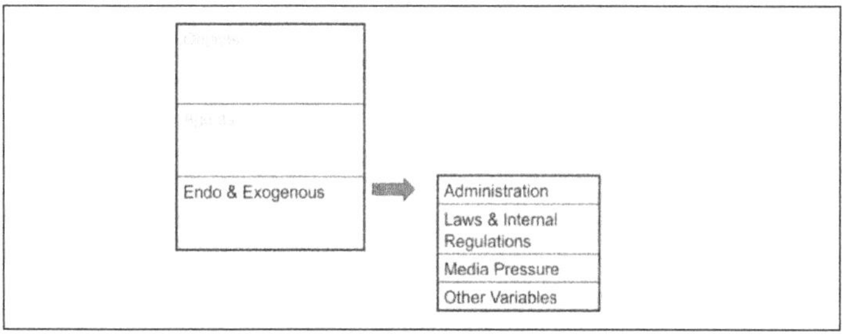

Figure 11: Endogenous and exogenous variables

External Agents, but their actions can influence decision makers. For example, policy makers might decide to cut funds and resources for museums, with important negative consequences to the adoption of ICT-related innovations – and without any possibility for the CHO to control or reverse the process. Moreover, the social networkers might generate online content that is unappreciated or might damage the museum's image, which could push the CHO to adopt corrective actions. On the other hand, online fans can largely contribute to the success of an exhibition: in terms of increased incomes, by providing favorable online feedback that encourages peers to visit the museum, at no costs for the organization. This phenomenon could motivate CHOs to invest more in social media.

In several examples, we have observed that all three dimensions of the framework dynamically affect each other. In the Objects dimension we noticed that Agents are dynamically involved in all the subcategories and phases, from the managerial definition of Goals and Expected Results, to the coordination of the planned Actions and analysis of results during the phase of Feedback. Agents also play a very important role in the third dimension of the framework: Endo-Exo variables.

THIRD DIMENSION: ENDOGENOUS AND EXOGENOUS VARIABLES

After having analyzed the first dimensions of the CHO2.1 framework (Objects) and the second (Agents), in this paragraph we'll focus on the third and last variable. The Third dimension of the framework consists of the Endogenous (internal) and Exogenous (external) variables (Fig.11) that might impact the CHO2.1 Framework. It includes four sub-categories: Administration, Laws and Internal Regulations, Media Pressure, and Other Variables .

When evaluating the possibility of adopting an ICT innovation to enhance the visitor's cultural experience, the analysis of Endogenous Variables might include, for example, internal administrative barriers; these barriers could affect other dimensions of the framework. For example, administrative constraints could impede the goal – under the Objects dimension – of selling online tickets for the exhibition via the internet due to the impossibility of using online financial procedures and providers not yet accredited. Other endogenous variables include Law and Internal Regulations, which might influence the adoption of innovative solutions. For example, internal regulations for public museums might delay the authorization to publish pictures of rare pieces of the collection via the web; such pieces, too fragile to be exposed to the public, need special care before digitization and becoming accessible online. Moreover, internal regulations requiring special permissions to proceed with the digitization could delay the Action Plan, and should be reported here.

On the other hand, examples of Exogenous Variables include legal aspects, such as the introduction of new laws that affect the CHO's investments. For example, new laws regulating internal security matters in preserving unique masterpieces could generate new expenses for the CHO and reduce the incomes of the budget allocated for ICT-based innovations.

The third subcategory of the third dimension of the framework is Media Pressure. In this case, for example, media could generate a successful donation campaign to open an online museum dedicated to children, affect visitor's attitudes and behaviors; criticize the CHO's production of social media contents; or stimulate a debate among policy makers around social media and the museum. These events could affect CHO decision makers and the adoption process of an ICT-based innovation. When applicable, these aspects should be reported in the CHO2.1 framework.

Finally, Other Variables – such as weather affecting open-air museums, or logistic impediments related to the preservation of the collection, including urgent intervention of conservation – could impact the (often) scarce CHO funds, and should be documented here. The subcategory Other Variables also includes unforeseen events that might affect the adoption of social media or other innovations. Some 5%-10% of the budget should be dedicated to the unforeseen.

The longitudinal CHO2.1 framework

The CHO2.1 framework follows a longitudinal perspective. In this respect, the three dimensions of the CHO2.1 framework dynamically affect each other over time. Let's imagine that, for health-related issues, a new – unforeseen – municipal legislation prohibits the use of the Wi-Fi technology in public buildings in the historical center of Paris. The new law will act as an exogenous barrier to the use of Wi-Fi technology for all the CHOs that are located downtown. If a CHO has already adopted the Wi-Fi solution to offer certain ICT services to its visitors, the new anti-Wi-Fi law will act as an exogenous variable that discards the ICT-based innovation already in use. In this case, the framework should be longitudinally updated and reviewed on the light of the new change. Several modifications will be expected affecting goals, targets, Wi-Fi-based applications and services. In this context, all three dimensions of the framework will be affected, particularly in terms of costs, legal aspects related to ongoing contracts, and resources. The CHO should regularly update the framework to reflect current and future scenarios, when single elements change over time. Moreover, if current conditions suggest discontinuance of an ICT-based solution, or if circumstances are not mature enough for the adoption of an innovation, it is important to verify if more favorable conditions might occur in the future, or alternative solutions are available at a different time.

Conclusions

This chapter has presented a new tridimensional longitudinal framework, CHO2.1, which is aimed at helping decision makers assess risk and opportunities, and offers a method to evaluate and adopt an ICT-based innovation.

The first question we have answered in this chapter is: are CHOs considering the potential risks associated with the use of social media? We have initially reviewed some of the possibilities, problems, and literature surrounding the adoption of social media for museums. The case for embracing or avoiding social media or ICT-based innovations has both advantages and caveats. Among the list of advantages is the fact that online interaction with the public might generate more interest towards an art exhibition, more visitors, and incomes. In addition, when applied to the educational context ICT solutions might further enhance the learning experience among online visitors, and generate viral interests among peers, at no cost for the CHO.

The second question we have addressed is: are cultural heritage organizations successfully exploiting the social media and ICT-based opportunities to deliver high quality services to users? In the cultural field we are in the infancy of the adoption of technology, and the successful use of social media or other ICT-based innovations from organizations is still a rare and a risky event for CHOs – particularly when management displays unfamiliarity with the ICT culture. There is an urgent need to help museums, libraries, and cultural organizations translate the excitement of social media and innovative solutions into practical reality and benefits. This might be done, only once proper conditions, methodology, planning, and resources allow.

At this point we wondered: are CHOs using a professional approach and method to evaluate if the organization is ready for Web 2.0 challenges? How can ICT-inexperienced CHOs treasure best practices, learn from previous mistakes, and find valid guidance for practitioners and internal staff who are taking the first steps into Web 2.0 innovations?

In this respect, the CHO2.1 framework has presented a new methodological approach that applies to ICT-based innovations for CHOs. After analyzing the first dimensions of the CHO2.1 framework (Objects) and the second (Agents), we focused on the third and last variable: Endogenous (Internal) and Exogenous (External) variables and their subcategories.

The framework can be applied whether the innovation is a new technology, a new mobile device for enhancing the educational experience, or a new way for visitors to access cultural objects through a technological application.

Today, social media is already history. What is new technology today, what is challenging the CHO today, will be different in five years. However, it is very likely that the next new thing will still challenge us. No matter what the next new thing in media studies will be, the CHO2.1 framework we propose here should provide CHOs with a solid intellectual and practical method and background to evaluate real necessities and available resources – and eventually adopt, in a stepwise fashion, ICT-based innovations.

Notes

1. Source: (Source: figure elaborated from Creative Commons http://commons.wikimedia.org/wiki/File:3D_Cartesian_coordinates.png)

References

Abbate, J .(1999) *Inventing the Internet*. Cambridge, MA:MIT Press.

Beniger, J.(1986) *Control Revolution:Technological and Economic Origins of the Information Society*. Cambridge,MA:Harvard University Press

Bijker, W., Hughes T. and T. Pinch (eds) (1987) *The Social Construction of Technological Systems: New Directions in the Sociology and History of Technology*. Cambridge, MA:MIT Press.

Castells, M., Fernandez-Ardèvol, M.; Linchuan Qiu, J. (2008). *Mobile comunication e trasformazione sociale*, Milano, Guerini e Associati, Italy.

Castells,M.(1996) The Rise of the Network Society. *The Information Age: Economy, Society and Culture, Vol.1*. Oxford:Blackwell.

Clark, W.; Logan, K.; Luckin, R. (2009). Beyond Web 2.0: mapping the technology landscapes of young learners, in *Journal of Computer Assisted Learning*, vol. 25, n. 1, pp. 56–69.

Gere, C. (2003). *Digital Culture*, London, Reaktion Books.

Falk, J. H (2009). *Identity and the Museum Visitor Experience*. Left Coast Press.

Falk, J. H., Dierking, L. D., and Foutz S. (2007). *In Principle, In Practice: Museums as Learning Institutions*. Altamira Press

Hansen, M.(2006) *New Philosophy for New Media*. Cambridge, MA: MIT Press.

Haythornthwaite, C., Kazmer, M.M. (2002). Learning, culture, and community in online education. *Research and Practice*,

New York, Peter Lang, 2002.

Haque, U. (2010) . The Social Media Bubble (blog), Guest edit series at *Harvard Business Review*. http://blogs.hbr.org/haque/2010/03/the_social_media_bubble.html Accesssed:Dec 12th, 2010

Jenkins, H.(2006) *Convergence Culture: Where Old and New Media Collide*. New York: New York University Press.

Jenkins, H., Purushotma, R., Clinton, K. et al. (2006). *Confronting the Challenges of Participatory Culture: Media Education for the 21st Century*, Chicago, MacArthur Foundation, 2006, p. 72. http://www.projectnml. org/files/working/NMLWhitePaper.pdf. Accessed: November 12, 2010.

Jones, C.; Dirckinck-Holmfeld, L."Analysing Networked Learning Practices" in Dirckinck-Holmfeld, L.; Jones, C.; Lindström, B. (eds.) Analysing Networked Learning Practices in Higher Education and Continuing Professional Development, *Technology-Enhanced Learning*, Sense Publishers, 2009, pp. 10-27.

Kelly L., Russo, A. (2008). *From Ladders of Participation to Networks of Participation: Social Media and Museum Audiences*. Conference paper. Proceedings of the Museum and the Web 2008 the international conference for culture and heritage on-line. Montreal, Quebec, Canada 09-12 April 2008. Jennifer Trant and David Bearman (eds.) pp.83-92.

Lévy, P. (1996). *L'intelligenza collettiva. Per un'antropologia del cyberspazio*, Milano, Feltrinelli, Italy

McLuhan, M. (1960). The Effects of the Improvement of

Communication Media. *Journal of Economic History* 20 (4) : 566–75.

Mor, N. (2009). Spatio-Tempo-Social: Learning from and about Humans with Social Media. Advances In Spatial And Temporal Databases. Lecture Notes, in *Computer Science*, Berlin, Springer, vol. 5644/2009, pp. 1-2.

Murero M., (2010). *Digital Literacy: Introduzione ai Social Media.* Libreria Universitaria, Italy.

O'Reilly, T. (2005). *What Is Web 2.0 – Design Patterns and Business Models for the Next Generation of Software* (blog). URL http:// oreillynet.com/pub/a/oreilly/ tim/news/2005/09/30/what-is-web-20.html . Access: Dec 15 2010.

Russo, A.,Watkins, J., Kelly, L., ,Chan, S. (2006). *How will social media affect museum communication?* Proceedings of the conference Nordic Digital Excellence in Museums (NODEM), 7th - 9th December 2006, Oslo, Norway.

Safko, L.; Brake, D.K. (2009). *The Social Media Bible*, New Jersey, Wiley.

Shih, C. (2002). *The Facebook era*. Prentice Hall Publications, Boston, Mass., USA, 2009.

Toumi, I. *Networks of innovation*, Oxford, Oxford University Press.

Wallace, P. (1999) *The psychology of the internet*. New York, Cambridge University Press.

Wellman, B.; Haythornthwaite, C. (2002). *The internet in everyday life*. Oxford, Blackwell Publishing.

Williamson, K.; Asla, T. (2009) "Information behavior of people in the fourth age: Implications for the conceptualization of information literacy" in *Library and Information Science Research*, vol. 31, n. 2, pp. 76-83.

Witmer, D.F. (2000) *Spinning the web*, New York, Longman, USA.

Xie, B. (2003). *Older adults, computers, and the Internet: future directions*, 2(4), pp.289-305. Accesso 20 novembre 2010, www.gerontechjournal.net.

Web resources

Assembly of the Museums Australia education group
http://maeg.textdriven.com

24hr Museum
http://www.24hourmuseum.org.uk/

The Panoscope 360: An Ethnography of the Appropriation of an Interactive and Immersive Technology in a Science Center

LETIZIA CARONIA
University of Bologna

ANDRÉ H CARON
University of Montréal

The Programme for International Student Assessment (PISA 2003 & 2006) reports a decreasing interest and performance of young people in sciences and technologies (OECD, 2003; 2007). This has generated considerable concern in many OECD countries. One of the main conclusions from the international conference on the *Declining Student Enrolment in Science & Technology* (Amsterdam, 14-15 November 2005) was that "OECD governments must take concrete steps to make science and technology studies more attractive" (OECD Global Science Forum and Netherlands Ministry for Education, Culture and Science, 2005). Although these results have been questioned (Lowell, & Salzman, 2007), the declining interest in science studies among young people remains a major issue for policy makers interested in maintaining high standards and traditional pre-eminence in these domains. Evidently, school and curricular activities are at the forefront, because they are the primary contexts in which young people supposedly acquire scientific knowledge and develop scientific interest and competences. However, schools are not the only learning context: extra-curricular activities in other settings can assist schools in developing interest in sciences with their own specific agenda, pedagogies, and related approaches to learning. The renewed interest in science centers' organization and their potential learning efficacy is based on the shared consensus of a need to develop and nurture young people's interest in sciences and technologies. At science centers, children and teenagers may discover these universes of knowledge in specific ways that may differ from those defined in curricular activities experienced in classrooms.

Most contemporary museums and science centers define themselves and are increasingly perceived as "places for learning" (Hein & Alexander, 1998) rather than as, for instance, places mainly intended to collect and preserve works of art, folk histories, archaeological discoveries, or specimens. Consequently, visitors are perceived as a special kind of client: they are "learners" (Hein, 1998). It then becomes of paramount interest to understand what kind of learning is or must be at the forefront.

The focus on what visitors learn and on how they learn within a museum context is strongly associated with this representation of museums as places where some form of knowledge acquisition takes place. A large body of research on museums and science centers has looked at short-term and long-term learning outcomes (i.e., knowledge and skills; perspective and awareness; motivation and interest; socially mediated learning (Falk, Scott, Dierking, et al., 2004; Paris, 2002). Contemporary research on learning in museums and science centers is also driven by a reference to the sociocultural approach to learning and development. Within this theoretical framework, learning in the museum is conceived as a kind of situated practice (Lave & Wenger, 1991) where social participation in object-mediated activities are the primary tools for developing competences and acquiring knowledge (Allen, 1997, Piscitelli & Weier, 2002). The focus of sociocultural research is to understand the processes through which visitors gain an understanding of the artefacts, installations, or exhibits.

Although social participation, social situated interaction, and the learners' ways of meaning-making are crucial dimensions in any learning experience, they are not the only factors at play. Bringing their own expectations, hermeneutics, and learning strategies, visitors situate the experience of particular artefacts within a specific context. The visitor-centered approach should be incorporated into "the complexity of the ecology of participation" (vom Lehn, Heath & Hindmarsh, 2005) which takes into account the multiple actors involved: the museum or science center, its inner organization, explicit mission, self-representation, and exhibitory strategies; the curators with their own specific views and folk theory of learning as well as time and budget constraints; the designers and engineers who construct the artefacts that will be exhibited according to their own specific framework; the artefacts' rhetoric, textual structure, and intended meaning. It is indeed within the horizon delineated by the text and the context that visitors display their strategies of sense making.

This paper presents the methodology and some results of an exploratory study conducted in a science center. The study analyzed an immersive-interactive technology, called the Panoscope 360, from its original design to its installation as part of a larger exhibit at the Montreal Science Centre. The principal goals of the study were to develop a formal model for the analysis of the context and the exhibit; to explore the phenomenology of the strategies adopted by the visitors to make sense of the artefact; and to evaluate the degree of fluidity of the artefact with respect to the visitors' strategies of sense making.

An ecological perspective was adopted which involved:

1. shadowing the progressive adjustments made to the artefact and the exhibit by the designers and the curators;
2. structurally analyzing the context and the implied educational model, the textual features of the artefact and its intended meaning, and
3. analyzing the visitors' naturally occurring ways of appropriating the artefact, its intended meaning, and the interactive learning experience it offered.

This paper focuses on the latter part of the study, which examined the visitor's appropriation of the artefact.

Artefacts as texts-in-context: a theoretical framework

From a semiotic perspective, artefacts are conceived as a text that has an intended meaning enacted in and expressed by its rhetorical devices (Eco, 1989; 1994). The artefact's features work to suggest certain meanings and courses of action rather than others. These intended meanings and forms to cope with the artefact are also projected by the co-text surrounding and accompanying the artefact. The labels, pictograms, warnings and instructions that are posted around the main text, as well as the history, explanation, and descriptions provided by various multimedia mobile devices such as PDAs (vom

Lehn, Heath & Hindmarsh, 2005; Woodruff, Szymanski, Aoki & Hurts, 2001; Woodruff, Aoki, Hurst & Szymanski, 2001), explain the artefact as a text of some kind (e.g., something to touch, something to watch, to manipulate, and to hear about) and link it to the outside community, namely to the institution and to the visitors.

Labels, legends, descriptions, history, and instruction for use are thus para-texts in the strict sense of the term (Genette, 1981, 1987). They encode what someone else stated about the artefacts. In so doing they link the organization's point of view and that of the visitors, and participate in creating the artefact's meaning and in framing the visitors' perspective. Para-texts work as an interpretive suggestion, and help establish a pact between the text and its consumers by indicating the realm of relevant encyclopedic knowledge consumers require to cope with the text (Lejeune, 1975; Eco, 1979). In the case of artefacts in a science center, the relevant background knowledge is generally constituted by naïve notions concerning physics, mechanics, biology, and the natural world, along with cognitive skills (observing, decoding, understanding, inference making). In this sense, these surrounding texts establish "a zone of transaction" (Genette, 1997:1) between the perspectives of the organization and of the visitors.

As is the case with all para-texts, the labels posted on the artefacts or the texts delivered through mobile devices may be ignored by the public, considered irrelevant, taken into account as a useful path for interpretation, or followed at face value (McManus, 1989). In any case, when acknowledged, these para-texts are interpreted by "real" people that make sense of them, debating and sometimes even defying their rationale.

Both the artefact and its co-text establish the hermeneutical horizon within which the visitors will accomplish their meaning-making process. Indeed artefacts and their para-texts may be conceived as "open works" (Eco,1989): they set the constraints and create possibilities for sense making, yet their intended and projected meanings and courses of action remain simply possibilities. The sense ultimately depends on the interpretive strategies of the visitors, their competences, expectancies and previous encyclopedic knowledge.

However, this process does not occur in a vacuum; it also depends on the context where it ultimately takes place: the science center and the exhibition room where the artefact is located.

To summarize, the meaning-making process depends on three crucial dimensions: the rhetoric of the object itself, the multilayered context where it is located, and the visitors' agenda. (Fig.1)

Understanding takes place at the intersection of these three dimensions and occurs as a situated and distributed process. It is situated because it depends on the local context and it is distributed because it depends on the different actors present, namely the artefacts' textual features and affordances, and the visitor.

This perspective has both theoretical and methodological

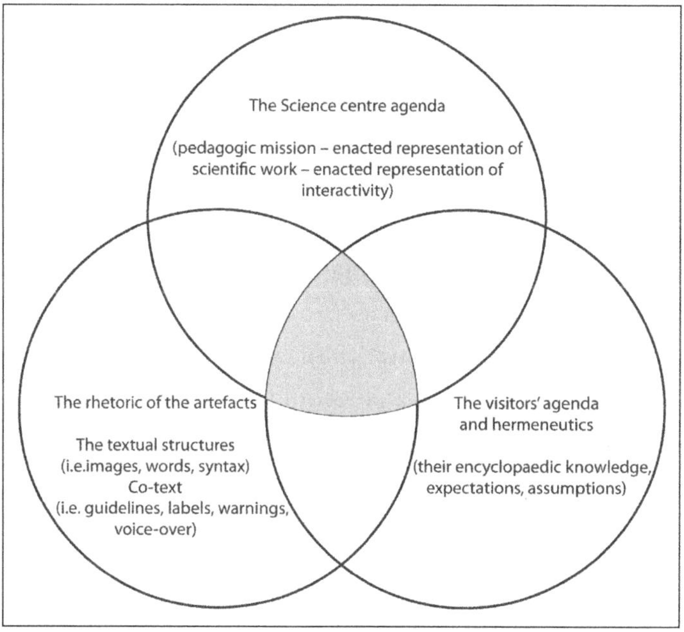

Figure 1. The emerging sense of artefacts

implications. The main theoretical implication of considering understanding as a situated and distributed process is that visiting an exhibit and experiencing an artefact are practices characterized by a high degree of indeterminacy regarding the outcome (i.e., whether the visit is a learning experience, what kind of learning is at stake, whether the visitors understand the meaning of the artefact). This degree of indeterminacy is not a measure of the failure of the artefact and/or the exhibitory strategies with respect to the intelligibility of the intended meaning. Rather, it is a characteristic of the way a contextualized open work challenges its audience. Consequently, research on learning in science centers should be process-oriented and focused on the multiple ways visitors make sense of the artefacts and create affordance-oriented and context-bounded meanings through local praxis.

Methodology: procedure and techniques

Conceiving the appropriation of an artefact for a science center as a situated meaning-making process calls for a methodological approach able to account for a diversity of instruments. To maximize the ecological validity of data, ethnographic methods and techniques were used, namely, non-participant observation of the natural flux of the visitors (notes on a grid), videotaping of the behavior inside the Panoscope through fixed cameras, audio-recording of naturally occurring verbal interactions during the visit with wireless microphones (families), and interviews with a casual sample of an important target audience of the science center (school groups).

Analyzing the visitors' agenda: Audio and audio-visual data are crucial for research interested in visitors' experience in museums (Gutwill, 2002). They allow researchers to analyze physical interactions with artefacts, movements in the interactive space, body language, and talk-in-interaction between co-visitors (Crowley & Callanan, 1998, Leinhard, Crowley & Knutson, 202, vom Leh, Heath & Hindmarsch, 2002).

Families and school groups visiting the exhibit and experiencing the Panoscope were recorded with fixed video cameras running consecutively during the opening hours of the exhibit on weekdays when school groups had booked a visit to the Centre, and on randomly selected weekend days, when families usually visited the science center. A total of 20 hours were recorded. Appropriate signs were posted to inform visitors that they were entering an area where video recording was present. The purpose of the recording was clearly explained, along with its use exclusively for academic purposes. In some instances randomly selected visitors (parents or children accompanied by parents) were asked if they were willing to participate in the study by wearing a wireless microphone during their visit. This allowed us to record their comments and conversations throughout the visit. Data on visitors' behavior were gathered during four subsequent periods from November 2007 to August 2008.

Data were analyzed according to a data-driven, inductive approach to category construction. Data were submitted initially to open coding (Charmaz, 2006) to find broad categories (i.e., observing without interacting, establishing play routines, making inferences, ignoring the running text). Second, data were analyzed again in a more detailed way to discern subcategories (i.e., courses of action such as touching the screen, taking pictures, walking through, guessing, or asking for the meaning of an image). In keeping with an inductive coding approach, new categories and related subcategories emerged as the analysis of data proceeded. The first data analyzed were thus recoded according to the new categories that emerged. The system of analytical categories was completed when saturation (or redundancy) was attained with respect to the categories developed to analyze the visitors' behavior. A fifth session of fieldwork was done in May 2009 to test the degree of saturation of the categories related to visitors' behavior data.

The following sections present a description of the artefact and of its running text and an analysis of the typology of visitors' forms of appropriation.

The Panoscope 360°: a structural analysis of the artefact, the text and its co-text

As described by Luc Courchesne, its inventor:

> *The Panoscope 360° is a horizontally immersive projection system using a single image channel. It is capable of displaying a full 360° x 180° hemispheric stereoscopic image from a single data channel. A hemispheric projector is placed above a downwardly flaring hemispheric screen; it projects an anamorphic disc image composed so the full horizon is placed at about 4/5 of the image radius. From within the installation, visitors see the horizon at eye level all around them, and are immersed in a distortion-free projected space. The image area holds within 95° vertically (15° above the horizon and 80° under) and 360° horizontally (personal communication).*

The content presented in the Panoscope 360 is a linear visual narrative about travel in the human body. Only the title of the story (*I can travel in the human body*) is presented in the beginning. Basically, the story recounts an exploration or voyage in the human body where the purpose is to resuscitate the heart, which has gone into cardiac arrest. The implied visitor begins the exploration through the eye, and travels through vascular vessels to reach the heart. This is where most of the story unfolds: the heart suddenly stops beating and the visitor must interact with various hand symbols that appear, and when touched, allow users to resuscitate the heart. An electrocardiogram line allows visitors to see if their actions are successful or not. After that, the traveller exits the body the same way as he entered it. The whole story lasts 2 minutes and 50 seconds, and unfolds with sound and images. There are no off-screen voices or dialogues. This text follows a classical narrative structure (context – situation – problem to resolve – solution – ending). It can be subdivided into four parts based on the images, sound, and rhythm of the content:

- 1st sequence: context (50 seconds) – title, representation of an eye with eyelids and iris.
- 2nd sequence: situation (20 seconds) – acceleration of images, globules, perception of travelling down vessels.
- 3rd sequence: problem to resolve and solution (70 seconds) – electrocardiogram line appearing, heart muscle pumping, interactive hand symbols appearing randomly.
- 4th sequence: ending (30 seconds) – the images become blurry as one travels back through the vessels to exit through the eye.

The images chosen to illustrate this voyage in the human body and tell this story are not part of the usual iconic representations used when presenting such information to children. Except for the eye illustration at the very beginning, the other representations of body organs (the heart and the vessels) are not represented by commonly used clichés (e.g., V-shaped heart). The images are therefore not self-explanatory and show a rather high level of non-definition. The soundtrack provides few interpretative cues. Typically, the audio effects give an indication of the heartbeat, its stopping and restarting, synchronized with the electrocardiogram and the narrative of the story.

The Panoscope 360 thus proposes a text which has an intended meaning that cannot be modified or completed by the visitors. The learning experience provided by the Panoscope mostly consists of understanding a story by grasping the immersive quality of the artefact, decoding images and sounds, and reacting to the interactive sequence of the running text.

A formal model for analyzing the artefact-in-context.

As stated above, the sense-making process of an artefact depends on three crucial dimensions: the context (i.e., the science center's pedagogic mission); the rhetoric of the artefact in terms of formal features (i.e., images, words, syntax) and co-text suggestions (guidelines, labels, warnings, voice-over); and the visitors' agenda and hermeneutic strategies.

To understand this meaning-making process it is necessary

to articulate both the *intentio operis*, which is the meaning inscribed in the structural organization and the affordances of the artefact, and the visitors' interpretive cooperation.

The artefact inside the science center may be classified along two main dimensions: the pedagogical model and the representation of interactivity enacted in its formal features and organizational structure.

The first dimension defines the learning goals at stake and may be conceived as a *continuum* ranging from:

1. Transmitting a corpus of already established knowledge; to
2. Transmitting to the visitors a sense of their agency in constructing knowledge (Rowe, 2002).

The second dimension refers to contemporary science centers making explicit reference to interaction as a principle organizing the visitors' experience with the artefacts (Adams, Luke & Moussouri, 2004). However, there is more than one way to conceive interaction that may be represented along a continuum ranging from:

1. Interactivity as a solitary relationship between the visitor and the object; to
2. Interactivity as social participation mediated by a semiotic artefact. (Heath & vom Lehn, 2008).

The two continua constitute the coordinates of the following formal model for a structural analysis of an artefact in a science center. (Fig.2)

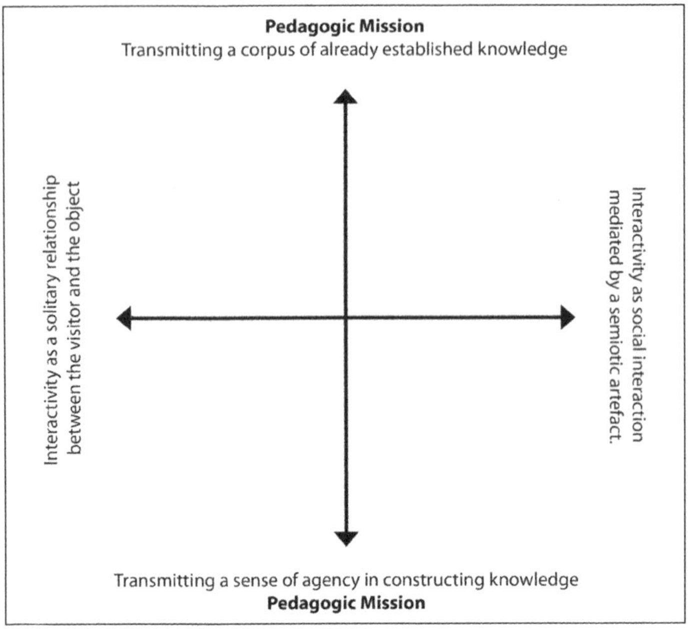

Figure 2: A formal model for the structural analysis of an artefact.

The vertical axis represents the scalar dimension of the educational model inscribed in the structural organization of an artefact. The extreme prototypes of this continuum are: the transmission of a corpus of established knowledge and the transfer of agency to the visitor.

The horizontal axis represents the idea of interaction inscribed in artefacts. The extreme prototypes of such a scalar dimension are the individual-object relationship and the social interaction mediated by the semiotic artefact.

The degree of an artefact along the two scalar dimensions determines where an artefact is situated in one of the four quadrants. On a larger scale, this model allows a simple

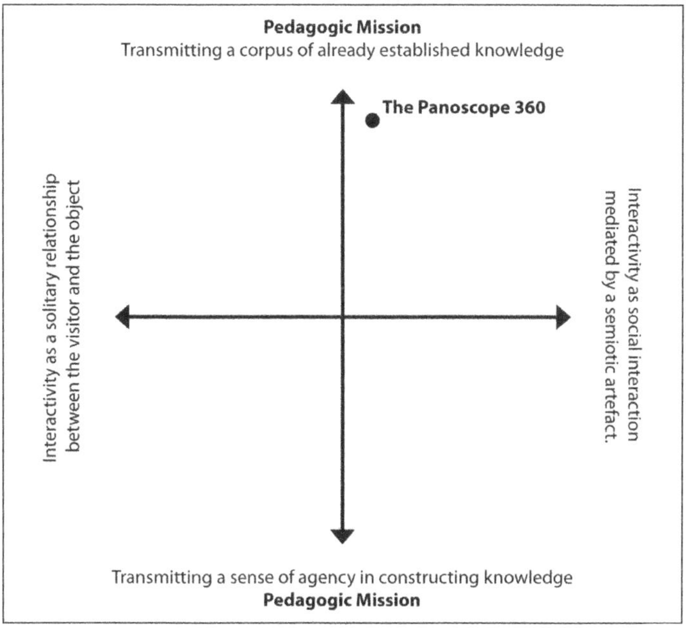

Figure 3: A positional representation of the artefact.

classification of a science center as such: the genre will depend on how many artefacts are situated in which quadrants.

Where was the Panoscope 360 located?

As mentioned earlier, the Panoscope 360 proposes a text that has an intended meaning that cannot be modified or completed by the visitors. Even if the artefact does not refer to the past or the history of scientific and technological discovery, it tells a story whose meaning the visitor needs to discover rather than to create. The Panoscope mostly implies the model of visitor that is a *reader/watcher*, whose experience of the object is mostly oriented to validate or discover knowledge, hypotheses, and ideas that already exist or are already formulated.

This cognitive-oriented organization is consistent with the exhibit's intended purpose. Often stated and even confirmed by the curators during the organizing meetings, the purpose of the exhibit was to bridge the distance between the visitors and the corpus of knowledge, ideas, and projects belonging to the selected circle of those working at the frontier of science and technologies. The exhibit was never meant to be an atelier or workshop but rather a presentation of the new and unknown territories of sciences and technologies. In this territory the visitor is supposed to learn by discovering what has already been discovered, but also by listening and watching what other people have already done or discovered.

It is then highly positioned with respect to the category *transmission of a corpus of an already established knowledge*.

The Panoscope is also mildly ranked with respect to the category *collective construction of meaning and social participation*. Even if it does not require social interaction or joint action to accomplish the task, its size, hemispheric structure, and the instructions provided by the co-text allow the simultaneous presence of more than one visitor.

This analysis based on the formal features of the artefact helps to define the intended meaning of the artefact and the learning experience inscribed in and projected by the textual organization of the Panoscope. It does not take into account what the visitors make of them.

An analysis of the visitors' forms of appropriation can reveal how these intended meaning and learning experiences may be confirmed, partially bypassed, defied, and even reversed by the visitors. Presented with the strategies displayed by the artefacts, the users apply their own tactics (De Certau, 1980).

The visitors' agenda

The goal of the ethnography of the visitors' behavior was to describe how visitors respond to the Panoscope and to its intended meaning during their visit to the exhibit. The aim was to delineate *the phenomenology of the natural ways of experiencing* the Panoscope. *Natural ways* means the repertoire of the types of interaction occurring in a situation that – although culturally and socially organized – is not manipulated by the researcher. Within this exploratory study, the objective was neither to quantify the occurrences of each type nor to measure which type of interaction was most frequent. Rather, the goal was to reconstruct the range of possible ways of experiencing the artefact.

The unit of analysis was defined as the time a visitor or a group of visitors spent inside the Panoscope from entrance to exit. The analysis of situated verbal (exclamations, comments, verbal exchanges) and non-verbal behaviors (physical posture, actions, gestures) served to create a typology of the forms of appropriation.

Form of appropriation refers to the relationship established by the visitor(s) with the artefact. A form of appropriation is characterized by dominant courses of action that are considered typical of a given form of appropriation.

According to the dominant observable behaviors, each unit of analysis may be classified as an occurrence of a single type or constituted of more than one form of appropriation (i.e., some visitors' evolving from text processing to play routine). Thus, the forms of participation are not necessarily mutually exclusive within the same unit of analysis.

Seven forms of appropriation were identified: walking in and out, uninvolved observation, pure interactive approach, pure entertainment, play routine, processing and expanding the text, and convergent-narrative approach.

Although the forms of appropriation indicate visitors' ways of making sense of the artefact, they cannot be considered pure cognitive stances. From a socio-constructivist perspective on learning and cognitive development, cognitive stances should be considered as related to both the semiotic artefact and the social participation structure of the experience in which the visitors are involved (Vygotsky, 1978, Cole, 1991). Whether the visitor explores an object alone or with some co-visitors may have an impact on the ways they make sense of the object.

The analysis of the visitors' behaviors considered two main contextual dimensions: the sequence(s) of the running text and whether the visitors were alone.

Here are a few examples of the behaviors classified as occurrences of the various forms of appropriation. Each example is contextualized according to the narrative sequence of the text corresponding to the observed behavior and to the social participation structure.

Walking in and out

Example 1

Running text: first sequence after the title

Social participation structure: a dyad constituted of a boy (estimated age: 8-10) and a girl, (estimated age 12-14)

A boy comes in followed by an older girl. They cross the space. They leave. They enter again. They observe the whole structure of the artefact. The girl pulls the boy by the sleeve. They leave. They do not talk.

Example 2

Running text: first sequence. Social participation structure: three girls (estimated age: 8-10).

They come in, cross the space, leave, come back again, and leave again. They do not have any verbal interaction.

Uninvolved observation

Example 3

2nd sequence

A single male adult

He comes in, observes the screen, crosses the space, looks at the line illustrating the cardiograms, and leaves.

In these kinds of appropriation, the visitors cross the space delineated by the structure, and look at the surrounding screen, yet they do not appear to be interested in the artefact or motivated to explore or discover its meaning. The visitors may also observe for a while, but they do not interact with the artefact or indicate that they grasped the intended immersive nature of the experience.

Convergent-narrative approach

Visitors adopting the convergent-narrative approach act in ways that are consistent with and sequentially related to the rhetoric and narrative structure of the running text (beginning – setting – introducing a problematic event – asking for visitors' interaction – resolution – ending). The visitors' stances are complementary to the intended meaning.

Example 4

1st sequence, a girl (estimated age: 10)

She observes; she touches the screen; she looks in front of her then turns to see the screen behind her.

2nd sequence

She looks at the whole screen surrounding her; she stretches her arms as if she wanted to be ready to react to something.

3rd sequence

A blue icon blows up; she looks surprised. She looks for new icons, rapidly turning around as if she understands that the icons for the interactive sequence may appear anywhere on the screen. She coordinates her movements and enters the interactive sequence of the text appropriately: sound, icon, visitor's action, text reaction.

4th sequence

She observes, she turns to watch the screen behind her, she leaves when the opening title reappears.

In this case of convergent-narrative approach, the visitor is alone. She follows the narrative, adopting behaviors that are complementary to the sequences of the story: at the beginning she observes, she interacts during the interactive sequence, she waits for it to end. Her body movements toward the screen (touching) reflect her understanding that she is supposed to do something to the artefact. Clearly, some previous knowledge about what the visitor is supposed to do is at stake and supports her making sense of the artefact. The behavior also reflects an understanding of the immersive nature of the experience: the body movements inside the structure indicate that the visitor grasped the specific perceptual characteristic of the artefact. The frontal vision is not sufficient to perceive the whole image and – as in a natural landscape – she needs to change her perspective to perceive the whole of it.

The convergent-narrative approach also takes place when more than one visitor is involved. (Fig.4)

Although basic, this kind of interaction with the text is an example of the convergent-narrative form of appropriation. The social interaction among the participants follows the narrative development and leads to shared inferences and understanding: what the experience is about, what the images represent, how the artefact works (actually it doesn't react to the shadow but to the movement), when it ends, etc. In this case, the text becomes the focus of joint attention, and the verbal interaction among co-visitors facilitates collective

1st sequence	
Adult:	this is the eye
Child: ((reading the title aloud))	I can travel inside the body
2nd sequence	
Adult and Child	((they look at the images))
Child:	((goes out and tells an adult who is at the entrance)) I (can) travel inside the body ((he comes back))
3rd sequence	
Female Adult	((she interacts with the artefacts, she puts her hands on the blue icons)). ((observes))
4th sequence	
Female adult	It works with the shadow (...) Here we come back to the eye (..) Here we go out

Figure 4: Dyad: female adult and child, (estimated age: 8-10).

understanding of the text.

In a convergent-narrative approach, the visitors' ways of acting are consistent with and sequentially related to the rhetoric and narrative structure of the running text (beginning – setting – introducing a problematic event – asking for visitors' interaction – resolution – ending). The visitor's stance is complementary to the intended meaning.

As the next example shows, this form of appropriation

may be much more sophisticated and may turn into a collaborative act of processing and expanding the text.

Processing and expanding the text

Visitors adopting the *processing-expanding the text approach* link what they see and do in the artefact to some previous experience or background knowledge. Some items become topics of conversation, whereas some topics give rise to hypothesis generation and inference making. (Fig.5)

The following sequences are embedded in a larger conversation occurring between two parents (father and mother) and their daughter (Chloé, age 12) during their experience inside the Panoscope. All the conversations were recorded through a radio microphone worn by the mother.

The participants follow the narrative sequence of the running text, they decode the images (10–12), and wonder what they mean (13–14), the adults scaffold their daughter's understanding, validate it (26–29), or repair some wrong inferences (40–41). Within this convergent-narrative collective form of appropriation, there also cases where visitors process the text according to some background encyclopedic knowledge (31–34) and expand it towards a new understanding, i.e., the platelets help in fighting microbes (15–17); the shape of a heart is not like the currently known icon of a heart (42–44). They briefly undertake the interpretive cooperation required by the text by elaborating on the images, questioning, expanding the information, and adopting a circumstantial evidence-based

1st sequence: after the title

10. Mother: we are like in a vein [in an artery
11. Father: [yeah some blood vessels
12. Chloé: in the – in the vein stuff, here in the eye
13. Mother: Hey Chloé look! We're like in a canal
14. Father: (???) yeah that's all the red globules
15. Mother: the platelets
16. Chloé: What are the platelets?
17. Mother: They are, they are like in our blood they help us to fight the microbes, otherwise there is no balance in the blood

[...]

3rd sequence

26. Mother: Do you hear?
27. Father: Yeah
28. Mother: Our heart has stopped?
29. Chloé: Beating
30. Mother: Yeah
31. Father: We give some jolts and it starts again
32. Mother: Oh!
33. Mother: Look! When dad gives some jolts to people that had a heart attack
34. Father: That's it here they are giving a heart massage. This is the valve which(??) the valve which(??)
35. Chloé: Yeah it's dead
36. Mother: Yeah but we stimulate the heart
37. Father: Because look!do you see it? there there is a: [a:
38. Mother: [yea the diagram
39. Father: A diagram
40. Chloé: Oh but look here there is one here there is one they say is alive
41. Mother: No this is where to put hands to make a a cardio resuscitation here you see the heart when we stimulate it by a heart massage
42. Chloé: It looks like this a heart?
43. Mother: Yea ehm it's with the ventricles
44. Chloé: I thought it had a heart shape

Figure 5: Processing and expanding the text.

method of interpretation. They thus use the text as a shared basis for further learning.

Purely interactive approach

Although the forms of appropriation follow the narrative sequence of the text (observation followed by interaction), the dominant approach is purely interactive. (Fig.6) The adults scaffold the encounter of the boy with the artefact, but only when and where some interaction is required. Their scaffolding is merely pragmatic. The focus of interest and joint attention is the hands-on characteristic of the artefact. There is no cue signalling an understanding of the narrative nature of the artefacts and of the whole story proposed by the text, which is appreciated only as an interactive machine.

Visitors showing this form of appropriation interact with the artefact either in accordance with the interactive sequence of the text (as in the example) or undertaking a sequence of hands-on behaviors without considering the feedback. In any case they act as if they ignored the global narrative meaning of the running text.

Play routine

In the case of a play routine the visitors establish social interactions, namely, play routines that are encouraged by or related to the running text. The issue at stake is mainly a social one, that is, interactions less oriented in understanding the text than creating or confirming the social links between co-visitors. The text is used as a pretext for doing something else.

1st and 2nd sequences: They observe.	
3rd sequence. They observe.	
Boy:	°oh!°
Adult 1:	(???)
Boy:	((he runs toward the screen trying to touch the icons. He pushes A2))
Adult 2:	It's okay
Adult 1:	There there you missed one
(5 sec.)	
Adult 2:	There!
Adult 1:	Again
Boy:	((he turns around trying to catch the blue hands)
9. Adult 2:	This way
10. Adult 1:	You're gonna (be hurt)
11. Adult 2:	Oh look there Joel! Move this way you're gonna be faster
12. Adult 1:	THERE THERE THERE! THERE! (.) THERE!
13. Adult 2:	THERE
14. Adult 1:	THERE!THERE! (just have) to make that up

Figure 6: Visitors: 2 adult females, a boy (estimated age: 8)

The example below (Fig.7) shows how the text and some information of the co-text are strategically used to play a game whose goals are establishing the leadership of one boy over the other and excluding the girl from joining the boys' group.

The dominant approach of this group of young visitors is clearly a purely interactive one. Although one boy reads the title aloud (2), the whole interaction with the artefact focuses on its interactive features (touching the screen when icons are on). There are no cues showing that the visitors grasp or follow the narrative structure of the running text: the collective inference making is mostly limited to decoding icons as if they were single pieces rather than components linked by a syntactic structure (12, 56, 57, 58). Although some statements signal that the visitors have grasped the immersive nature of the perceptual experience, they do not build on this. Rather, they make collective inferences on how the artefact works and on what they are supposed to do inside it (4–6; 7, 8, 16, 24-25).

In this case, however, the purely interactive approach with the artefact is strongly mediated by the social relationship and has dramatic social meaning and outcomes.

The young visitors seize the opportunity to take advantage of the running text, the physical shape of the artefact, and the instructions provided by the co-text to establish a play routine. They transform the instruction, setting the limit of three people inside the Panoscope in a play rule that involves gate-keeping (9, 22, 38). By routinely and strategically evoking this norm, they create a closed interactive space where some are admitted and some are excluded (41, 44, 50). The negotiations that take place ("you don't have the right", "yes I have the right," "ok, I'm going out", see turns 13, 14, 15, 19, 43, 49, 51) are typical examples of peer culture routines and ways of establishing a social organization.

What is typical of this form of appropriation is that although the visitors' verbal exchanges and the actions taken within the Panoscope are text-related, the objective at stake is eminently social. The objective obviously changes

1st sequence: The title is on	
1. Boy 4:	Touch it ((he touches the screen showing what to do))
2. Girl:	I can travel in the human body ((singing aloud the title))
3. Boy 2:	((sings)) Hey! (???)
4. Girl:	How does it work?
5. Boy 1:	It may be something you need to stay in the middle
6. All:	((they go in the middle of the Panoscope))
7. Boy 4:	as I did before. Touch it touch it
8. Boy 4:	The goal is touching the hands
1st sequence: The icon of the eye is on the screen	
9. Boy 4:	It's maximum three people!
10. Boy 4:	Ok more again
11. Boy 2:	[Hey I am the third
12. Girl:	[LOOK! THE EYE!! ((she goes toward the screen, she brings her body closer to the wall))
13. Boy 4:	I got here first
14. Boy 3:	I was here second
15. Boy 2:	Third
16. Boy 4	You need to wait for the hands
17. Boy 3:	Is it really maximum three people?
18. Boy 2:	They are beautiful
19. Girl:	Okay I'm going out
20.	((the girl and boy1 get out. Two more boys are coming in))
21. Boy 2:	[Here they are four (.) five (.) six!
22. Boy 4:	[It's maximum three people. It's maximum three people
23. Boy 3:	Are we three now?
24. Boy 2:	What are we supposed to do?
25. Boy 4:	the hands will soon arrive. It's maximum three people
26. Boy 2:	Okay where are the hands? ((irritated))
27. Boy 1:	Come on!
28. Boy 2:	What's that
29. Boy 4:	They are coming soon
30. Boy 4:	It doesn't work there are no hands ((he moves toward the exit yet remains at the entrance looking at the other boys still inside and others coming in again))
[....]	

31. Girl:	((staying at the entrance)): what are you doing?
32. Boy4	We're going to touch some hands (.) anyway It's maximum three people. Oh yes! the hands will come I remember this part
33.	3d sequence
34. Girl:	((remaining on the entrance)) HEY!THERE THERE HERE ((indicating the screen))
35. 3 Boys:	((they touch the hand))
36. Girl:	((always from the side))I will check this point
37. Boy1 [	Ah Remi
38. Boy4:	[No, It's It's maximum three people
39. Boy 2:	You guys you check there I check here
40. Girl:	((she touches the screen always staying on the side)) Once more
41. Boy4:	Roxanne! Its maximum three people
42.	((a fourth boy comes in))
43. Girl:	It doesn't matter
44. Boy4:	And you don't have the right to participate if it is only three people
45. Girl:	((she touches a hand while remaining on the side))
46. Boy4:	Its three people max
47. Girl:	I know=I know
48. Boy3:	THERE!
49. Boy5:	It never been three persons maximum
50. Boy4:	But you don't have the right
51. Boy5:	Yes I have the right
52. Boy3:	Euurk!ouak!!
53. Boy 5:	Look it will soon start again and I will have the right
54. Girl:	Its almost finished
55. Boy 5:	Yeah
56. Boy 3:	Look I'm in a kind of a giant whirl
	Some boys get out
57.Boy 5:	Look this is the brain
58 Boy 4:	Yeah I'm in the brain
59.Boy 5:	Shall we start again?
60.Boy 4:	Yeah its fun
	Adult walks in
	Let's go on the next turn

Figure 7: Participants: 4 boys and 1 girl, same school group (age 10) came in

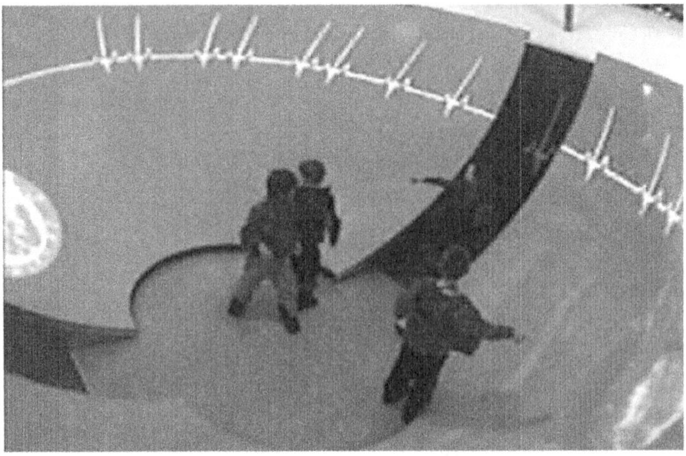

Figure 8: Play routine and social purposed interaction.

given the average age of the visitors and their gender (i.e., a group of adolescent boys making some deliberately obscene interpretations of the images on the screen and dirty jokes; girls showing theatrical disgust), yet the basic process remains: the artefact becomes an interactive field and the running text presents opportunities that can be integrated in a social performance. (Fig.8)

Pure entertainment

The visitors transform the artefact into something else by using the size, the form, and the material structure of the artefact as an interactive space or a playground. (Fig.9) A key difference with the play routine is that interactions occurring with the artefact or with co-visitors are neither caused by nor linked to the ongoing text. As the illustration below shows, visitors use the structure as a space for doing

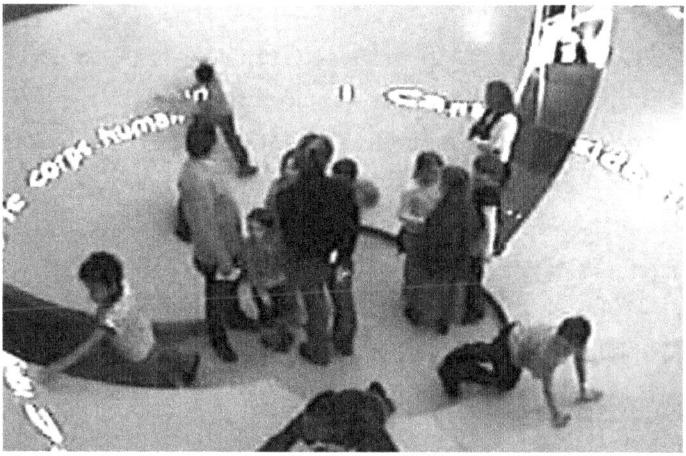

Figure 9: Pure entertainment.

things: running, climbing the wall, sliding down; teenagers also take pictures of themselves or simply chat.

Making sense of the artefact

The different forms of appropriation may be seen as resulting from a negotiation between the artefact, its co-text and context, and the visitors who adopt their own stance toward the meaning inscribed in the artefact. In this sense, the forms of appropriation may be distributed along a continuum showing the kind of relationship established by the visitors with the inscribed meaning. The extreme poles of this continuum are noticing and understanding the implied meaning and ignoring it. The two poles neither correspond to any normative taxonomy nor do they imply an underlying pedagogical assessment. They indicate the range of the possible ways of making sense of the artefact and its intended meaning.

The table (Fig. 10) illustrates the typology of the forms of appropriation of the Panoscope 360 observed and how these forms of appropriation reflect the interest in the artefact as well as the visitors' stance toward its intended meaning

The typology of forms of appropriation indicates the artefact's degree of fluidity regarding what can be made of it. Although the Panoscope has an intended meaning and provides the visitors with all the necessary cues to infer this meaning, it lends itself to uses and re-making that are not necessarily related to noticing and understanding this meaning. It appears to be a fluid artefact that allows for divergent sense making and may be appropriated in different and not mutually exclusive ways.

Some conclusions

According to the structural analysis, the Panoscope 360 may be located in the upper right part of the quadrant which combines the pedagogic mission and its representation of interactivity implied in an artefact. The visitors phenomenologically displace the Panoscope from the upper right part of the quadrant to the lower right part.

Designed to tell a story about traveling in the human body from the eye to the heart and to transfer knowledge by an immersive experience, the Panoscope also becomes an artefact that transmits a sense of agency. Proposed as an experience that may involve more than one visitor but that does not require social interaction, the Panoscope becomes

Forms of appropriation	Distinctive courses of action	Degree of interest	Visitors' stances toward the inscribed meaning
Walking in and out	Come in, have a quick look, go out	Low	Ignoring the inscribed meaning. ▲
Uninvolved observation	Gaze toward the screen, frontal vision, static posture, the visitor leaves without establishing any contact other than visual.	Low	
Pure entertainment	Skateboarding on the surface; taking photos; running over someone, sitting or laying down.	High	
Play routine	Establishing a competition (i.e., girls vs. boys; who is the fastest in putting hands on), showing leadership by telling others what to do.	High	
Purely interactive approach	Hands on the icons, touching the screen, grasping and participating in the stimulus-response-feedback sequence. The visitor goes out when this sequence ends or when there is no feedback.	High	
Processing and expanding the text	Referring to some notions, making some text-extra-text links; formulating questions about the possible meaning; trying to decode images.	High	
Convergent-narrative approach	The visitors observe the beginning, wait watching the screen, interact during the interactive sequences, observe. At the end of the running text they leave. They change attitudes according to the narrative evolution.	High	▼ Noticing and understanding the inscribed meaning.

Figure 10: The typology of the forms of appropriation.

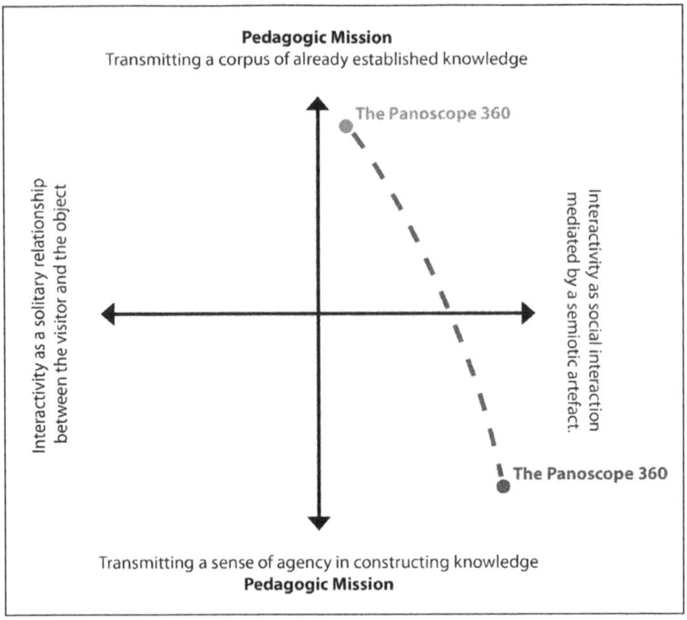

Figure 11: Displacement - the viewer's sense making of the artefact.

also a social context for shared understanding, cooperative learning, and social interactions. Visitors interpret the Panoscope material structure and textual cues in ways that diverge from those anticipated by the designers and the curators, and make it work as something else: a cognitive space is transformed into a social space.

What allows such a repositioning?

Aside from idiosyncratic variables (such as age, previous knowledge of the topic at hand, etc.), this displacement, or at least the possibility of such a displacement, seems to be related to three extra-individual factors:

- the context;

- the inferential quality of the text;
- whether or not the visitors acknowledge and use the information provided by the co-text.

The science center suggests a *learning by doing and by discovering* approach and therefore frames the visitors' general perspective: artefacts and exhibits inside this context are perceived as things that lend themselves to being manipulated. The context legitimizes a kind of approach not centered on following instructions. Within this context, the Panoscope acquires an aura of indeterminacy irrespective of its intended meaning and intended learning experience.

The text chosen to tell the story of travelling in the human body was an iconic one. Despite the rhetoric surrounding iconism, when images are not accompanied by words they are often unsaturated media: they require inference-making more than verbal descriptions. In this sense the Panoscope is a very "open work": it has a meaning, but its interpretation is not taken for granted; it depends on the visitors' "inferential walks" (Eco, 1994) within the field of possible meanings.

The semantic and pragmatic scaffolding provided by the co-text supposedly frames this interpretive path and establishes the boundaries of the visitors' sense making. Yet reading or hearing the co-text suggestions is not mandatory. As the data show, when the co-text is not acknowledged by the visitors the interpretive drift is much more significant.

Within this specific context, the Panoscope 360 seems to be a perfect fit for the two possible missions of a science center: transmission of a corpus of knowledge and transfer of a sense of agency in meaning creation. It proposes some knowledge to be inferred and discovered and provides a space for negotiation where the visitors feel free to create divergent meanings and transform the suggested learning experience into something else.

Whether this outcome is suitable is not an empirical question; it depends on the goals of the exhibit and the mission of the science center. This research has shed light on the process at stake and on the factors at play in decision making.

This research was made possible with the generous support and funding of the Bell Chair in Interdisciplinary Research on Emerging Technologies , the Canada Foundation for Innovation and the Centre des Sciences de Montréal.

References

Adams, M. Luke, J., and T. Moussouri. 2004. Interactivity: Moving Beyond Terminology. *Curator*, 47(2): 145-169.

Carugati, F., and P. Selleri. 1996. *Psicologia sociale dell'educazione*. Bologna: il Mulino,

Cole, M., 1991. Conclusion, *Perspectives on socially shared cognition*. edited by L.B Resnick, J.M. Levine, and S.D. Teasly, 398-417. Washington, DC: American Psychological Association,

Cole, M., and J. Wertsch. 1996 "Beyond the individual-social antinomy in discussions of Piaget and Vygotsky." *Human Development*. vol. 39, no 5: 250-256.

Crowley, K., and M. Callanan. 1998. "Describing and Supporting collaborative scientific thinking in parent-child interactions." *Journal of Museum Education*, 23(1):12-17.

Eco. U. *The open work*. Cambridge MA: Harvard University Press, 1989.

Eco, U. 1994. *Six walks in the fictional woods*. Cambridge MA, Harvard University Press.

Gutwill, J. 2002. Gaining Visitor Consent for Research: testing the Posted-Sign Method. *Curator*, 45(3): 232-238

Heath, C., and D. vom Lehn. 2008. Construing interactivity: enhancing engagement and new technologies in science centres and museums. *Social Studies of Science* 38(1): 63-91.

Heath, C., D. vom Lehn, J. Osborne. 2005. Interaction and Interactives: collaboration and participation with computer-based exhibits. *Public Understanding of Science* 14(1): 91-101

Hein, G.E. 1998. *Learning in the Museum*. London: Routledge.

Hein, GE, and M. Alexander. 1998. *Museums: Places of learning.* Washington, DC: American Association of Museums.

Leinhart, G., K. Crowley, and K. Knutson, eds. 2002. *Learning Conversations in Museums.* Mahwah, N.J.: Lawrebnce Erlbaum Associates.

Lowell, L. and H. Salzman. 2007. *Into the Eye of the Storm: Assessing the Evidence on Science and Engineering Education, Quality, and Workforce Demand.* The Urban Institute. Retrieved in December the 9th, 2009 from http://www.urban.org/publications/411562.html

McManus, P.M. 1989. Oh Yes They Do: How Museum Visitors Read Labels and Interact with the Exhibit Texts. *Curator,* 32(3): 174-189.

OECD. 2003. *Global Science Forum Activity on Declining Interest in Science Studies Among Young People.*" Retrieved from http://www.caos.nl

OECD Global Science Forum and Netherlands Ministry for Education, Culture and Science. 2005. Conference press release. *Declining Student Enrolment in Science & Technology: Is It Real? What Are The Causes? What Can Be Done?* Amsterdam Koepelkerk Convention Centre, The Netherlands.14-15 November 2005. Retrieved the 9th December 2009 from http://www.oecd.org/document/

OECD, PISA,2006. *Science Competences for Tomorrow's World.* Vol 1 & 2. Geneva: OECD Publishing, 2007.

Paris, S.J., ed. 2002. *Perspectives on Object-Centered Learning in Museums.* Mahwah, NJ: Erlbaum.

Piscitelli, B., and K. Weier. 2002. *Learning With, Through, and*

About Art: The role of Social Interaction. Perspectives on Object-Centered Learning in Museums, edited by S.J. Paris, 121-152. Mahwah, NJ: Erlbaum.

Resnick, L.B, J.M. Levine, and S.D. Teasley, eds. 2001. *Perspectives on socially shared cognition*. Washington: American Psychological Association.

Rowe, S. 2002. *The role of objects in active, distributed meaning-making. Perspectives on object-centered learning in museums*, edited by S. Paris, 19-35. Mahwah, NJ: Lawrence Erlbaum Associates.

Salomone, I. 1997. *Il setting pedagogico. Vincoli e possibilità per l'azione educativa*. Roma:Carocci.

Tramma, S. 2009. *Che cos'é l'educazione informale*. Roma:Carocci.

vom Lehn, D., C. Heath, and J. Hindmarsh. *Rethinking interactivity: design for participation in museums and galleries*. Work, Interaction & Technology Research Group, King's College London. (2005). Retrieved January 26, 2007, from Re-thinking Technologies in Museums at http://www.idc.ul.ie/museumworkshop/

vom Lehn, D., C. Heath, and J. Hindmarsh. 2002. Video-based field studies in museums and galleries. *Visitor Studies Today*. 5(3): 15-23.

Wenger, E. 1998.*Communities of practice: learning, meaning and identity*. Cambridge: Cambridge University Press.

Wertsch, J.V. 1991. *A sociocultural approach to socially shared cognition. Perspectives on Socially Shared Cognition*, edited by L.B. Resnick, J.M. Levine, and S.D. Teasley, 85-100. Washington, DC: American Psychological Association.

Wertsch, J.V. 1998. Mind as Action. New York: Oxford University Press.

Wertsch, J., P. Del Rio, A. Alvarez, eds. 1995. *Sociocultural studies of mind.* Cambridge: Cambridge University Press.

Woodruff, A., M.H. Szymanski, P.M. Aoki, and A. Hurst. *The Conversational Role of Electronic Guidebooks.* Lecture Notes in computer science Proceedings of the 3rd International Conf. on Ubiquitous Computing, edited by G.D. Abowd, B. Brumitt, and S.A. Shafer, 187-208. London, Springer Verlag 2001.

Woodruff, A. P.M. Aoki, A. Hurst, and M.H. Szymanski. *Electronic Guidebooks and Visitor Attention.* Proceedings of the 6th International Cultural Heritage Informatics Meeting (ICHIM 2001), 437-454. Milan, Italy, Sep. 2001.

Click History: Wherever, Whenever

KATHLEEN HULSER

New-York Historical Society

STEVE BULL

Cutlass Inc

How do you engage new audiences both inside museum walls and out in the cityscape with museum objects, library writings, paintings, and long-dead personalities? Cellphone tours and iTunesU content from a collaboration of the New-York Historical Society and Cutlass give some answers in a technological key. They include a glimpse of a Haitian grave robbed in Lower Manhattan to supply a saint for an uptown cathedral, and a teen rapping about the Kansas-Nebraska Act. Hulser and Bull have discovered that gaming Miss Grundy or a GPS *Where's Waldo?* may be the route to having history come alive in the devices that young hands best love to hold.

Museums realize that their holdings ought to be seen inserted into their original contexts and cityscapes. They also reluctantly recognize that many new audiences find words like *historical society* about as welcome as *dental surgery*. However, a game of discovery of the mysteries of the past in familiar cityscapes is a game that young and old love to play.

Museums in general own a plethora of objects that no one ever sees, and could never conceivably be presented in real exhibitions. However, as we well know, web appetites are truly gargantuan, and handheld devices actually work at a scale precisely that of many tiny, precious, rarely known objects. So the group of images, which correspond to a grave just spied through a crack in the fence of St. Patrick's Old Cathedral, are the small family portraits of "almost saint" Pierre Toussaint. The originals painted on ivory are only a little over three inches high. Thus on an iPod, iPhone, or similar smart phone found with every young adult, you are

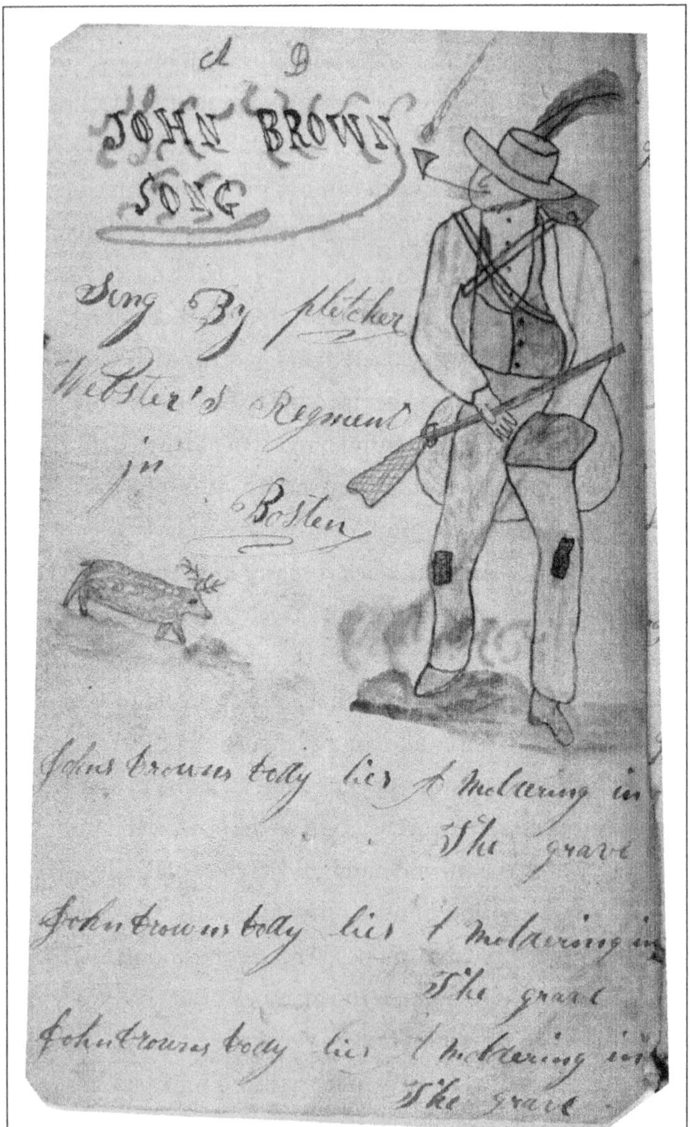

John Brown's Body Song by Fletcher Webster's Regiment. Civil War Songbook. Nov. 24, 1862. GLC00968. Courtesy of Gilder-Lehrman Collection.

seeing them at actual scale in the palm of your hand.

So while delighting curators by giving material culture a boost, we also intrigue the itchy finger public that must do something with their phones. Museums ought to be more than pleased that their long-term agenda of restoring context to objects is happening in such an unanticipated fashion. This internal museum agenda happily dovetails with the participatory urges of device users, who enjoy browsing for its own sake. The cellphone public thinks in terms of vignettes rather than large scale exhibitions. They often have more experience and love of collecting and collectibles than reverence for the treasures of the museum mentality. The words *search, find, sort, tag, save* may hold at least as much relevance as *learn, emulate, revere, and inspire*, although both may be active impulses at different times in the same user.

Meanwhile, the small device, and its parent, the computer, have radically revised assumptions about viewer attention. Although the customary museum wisdom has too long been a great sob that young people have short attention spans, few museums seem to be absorbing the fact that the museum visit now includes experiences that are before, during, and after an actual bodily walk-through in the museum. Our N-YHS iTunesU site has had 10,000 tracks of one walking tour downloaded in a year – an indication that we are only on the threshold of a huge wave of virtual use of various digital packages. So, at the New-York Historical Society, a young black man might be thrilled to learn about *Slavery in New York*, or he might dial a cellphone tour that tells him about spots

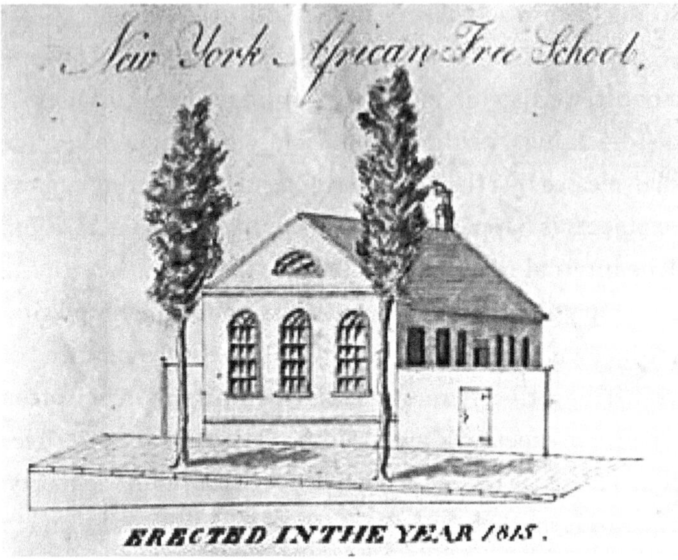

African Free School. Patrick Reason. Records of the New York Manumission Society. New-York Historical Society Library.

where fugitive slaves sheltered during the 1850s in Tribeca. Or a Latina student at the Borough of Manhattan Community College might be contemplating a sign on Chambers Street, noting that Frederick Douglass landed there before launching a big anti-slavery career.

These before-, during-, and after-visit experiences encourage the museum to handle its materials differently. At N-YHS we have explored the story of black political activist Dr. James McCune Smith from many angles. He was a star graduate of the African Free School. The N-YHS has the African Free School's records and can place his activity in the context of early segregated schools, which were cradles of black political thought and also a symbol of early Jim Crow in the North. We can suggest Underground Railroad activities in his pharmacy at 55 West Broadway with apothecary jars from our collection; this time the containers are not just pieces of material culture, but metaphorically signal remedies for injustice. The branching possibilities of media, the linkages between a cellphone tour and digitized collections, and full-fledged exhibition websites mean that both clickers and listeners have great latitude in exploring the story.

Once museum people begin to think of encounters with the collections as something happening in an unrestricted time and space – mobile devices, computers, and social media add up to the kind of visit extensions that should make audience evaluators drool.

Consider the following three scenarios:

1. A family with child in stroller strategizes about how they can enjoy some history without sparking too much protest from the youngest member by stopping. The solution is the peripatetic history experience, which allows mom and dad to keep in motion with a hand on the stroller bar, ear buds in, and an eyeball on images flowing on iPhone.
2. Another modern entrée into the historic cityscapes of a place saturated with history is the cellphone tour, which offers rare images of what once was, cued to a walk around neighborhoods that look different now. *Hidden Sites of Slavery and Freedom* did precisely this, animating a walk by old warehouses and trendy

restaurants in Tribeca, while talking of how black political thought was born in those very streets.

3. GPS technology now allows for prompts, so the historically minded can have themselves pinged to alert when passing by the very spot where David Ruggles ran a freedom reading room, which Underground Railroad activists used to get information on slave catchers and fugitives.

These shifts in visitor experiences and encounters with collections have many implications for 200-year-old institutions like the New-York Historical Society.

At a very practical level, it means that N-YHS has an incentive to photograph and digitize many items which might not otherwise be excavated for anyone's use. Usually, the N-YHS will photograph very important and valuable items, objects used in exhibitions, and objects requested by scholars or publications. That leaves the bulk of our collections without images, even non-professional reference photographs. Now that online demands are being made, justification for photography is there, and the rationales for photographing items have changed. So, we have extracted samples from anti-slavery pamphlets or abolitionist song books, and sketches in letters.

Our experience in recent years is revising scholarly assumptions about short form knowledge. Scholarship, and the traditional museum which functions as the art-historical branch of such scholarship, have emphasized long-form knowledge. Meanwhile, the icon fluency of the web has garnered credibility for cognitive theories that view knowledge as an accretion of small bits. Mobile devices and social media play to our fascination and impulse-driven appetites for short-form knowledge, which is extending museum experiences into before-and-after explorations, side paths, collaborations, virtual collecting, gaming, and urban exploring. Size no longer matters.

Our audiences are teaching us the pleasures of the collections by creating new things out of old, drawing on their fresh sensibilities. For example, a summer high school intern group created a teen tour of the exhibition *Lincoln in New York*.[1]

We feel sure that the chanted history poem created by Tonie McKenzie and Omosede Eholor is the only existing female rap on the consequences of the Kansas-Nebraska Act.[2]

Everyone these days talks about doing something with collections: we stand in awe of what these two groups – all under age 17 – actually did with ours. Further changes in our authoring template are under way to ensure that the choices of curators of the past will resonate with the audiences of tomorrow. We welcome the way electronics allow people to play in the collections, and trust good history will emerge in many forms.

These examples suggest that new media is changing the clock on the museum history experience by motivating people to sample the fare before and after, not just during a visit. They allow users to encounter historical things virtually on their own schedule, in small and large doses. As social media enables more participation, the future of history through technology will involve the increasing collaboration of users yet to be known.

Notes

1. Listen at: http://touchtonetours.com/audio/LINCOLN_112.wav

2. Tune in at: http://touchtonetours.com/audio/rap.wav.

Portable and Playful: Locative Media, Art and Games in the Mobile Space

MARTHA LADLY

Ontario College of Art & Design University

Mobile public art involves social practices that are playful, provocative, and by definition, ephemeral; locations are not fixed and audiences and players are on the move. Artists and designers mediate their works between traditional museum and gallery environments, virtual environments, and the built landscape or public space, which forms the *set* or *board* for their work. Audiences are not passive; they are participants and players, using social media and their mobile devices to interact with each other, the artworks and games, and the environment.

This paper will examine the rich history and questions posed by these hybrid forms found in mobile "locative" art-making practices and some of the challenges that artists, designers, and audiences face in creating and participating in locative art projects accessed by mobile devices in public space.

A number of pioneer artist investigators were laying the groundwork for mobile public art, using combinations of location awareness and mobile technologies. In 1991 Janet Cardiff created her first audio walk at the Banff Centre, in Alberta. *Forest Walk* guided participants with an audio cassette deck and headphones through the local pine forest, accompanied by walking directions and observations, interspersed with a dramatic audio dialogue. Cardiff admits that the instructions were difficult to follow and the sound quality wasn't great, but it changed the artist's thinking about the potential for this new form, and became the prototype for all of her later mobile audio walks and artworks.[1]

Masaki Fujihata, one of the pioneers of Japanese new media art, created an early locative work in 1994, using a car-mounted laptop computer equipped with GPS and a video camera to create a locative image map of Mount Fuji. His resulting artwork *Impressing Velocity* made the data and images available for viewers to explore interactively.[2]

In 1999 artist Teri Rueb launched *Trace*, one of the first geo-annotated mobile art projects, using GPS coordinates embedded in the landscape to access a database. Her interactive walk was a memorial environmental sound installation, created as a site-specific response to the network of hiking trails near the Burgess Shale fossil beds in Yoho National Park, British Columbia. This work was a real precursor to the phenomena of mobile public art: the participant carried a custom knapsack equipped with a portable computer, headphones, and a GPS receiver, and memorial poems, songs, and stories contributed by collaborators played in response to the participant's movements through the landscape, triggered by GPS coordinates.[3]

A recurring theme in these seminal mobile location-based artworks is their situation in remote environments, and places that you have to rely on your senses, your powers of observation, and your fitness to experience them. Participants using portable technologies were able to navigate landscape-based links to the artist's works, but only by also locating themselves in the artist's chosen, often difficult, and wildly beautiful locales. Mobile art gave artists and audiences – who are active participants – our bodies back, and expected us to be up to the physical challenges!

These artworks were also among the first to investigate the convergence of geographical place and data space, reversing the idea that digital art only existed in placeless virtual locations, usually accessed on computers via the Internet. Drew Hemment, a locative media scholar, remarks: "Locative art's focus on digital authoring within the environment, on a dynamic relationship between database and the world, offers the chance to take art out of the galleries and off the screen." [4]

Locative media artists and researchers found themselves facing profound and explicit inter-dimensionality in their artistic proposals, such as the ability to simultaneously move through and address physical space and electronic space. Their work explicitly acknowledged the potential of GPS, adapting it for geographically and socially mediated artworks and investigations.

Karlis Kalnins, web-artist, mapper, and media activist, was at the K@2 workshops in the summer of 2003. He termed their work "locative media" as a test category for processes that could facilitate the annotation of physical space, by combining mobile data communications with GPS and mobile computing. The term stuck. Its catalytic premise was the growing civilian awareness and engagement with a construct that had distinct military origins. Hosted by media art collective RIXC5 in the summer of 2003, an international network of researchers working with mobile, context-aware computing devices and applications held their first location-based workshops on the site of the former

Soviet military naval base of Karosta, on the west coast of Latvia.[6] Rasa Smite and Raitis Smits, Latvian artists who were working with the *Acoustic Space Lab* project at the RT:32 Radio Telescope in Karosta, led a series of transcultural mapping workshops to further explore the territory.[7] Locative artists and researchers found themselves facing profound and explicit interdimensionality in their artistic proposals, such as the ability to simultaneously move through and address physical space and electronic space. Their work explicitly acknowledged the potential of Global Positioning System (the satellite-based navigation technology developed by the US military), adapting it for geographically and socially mediated artworks and investigations. With the introduction of GPS and mobile technology into the public realm, they saw new possibilities for social interaction and "a way to reclaim public space as a site for a new kind of shared experience."[8]

Most mobile artists and game designers find their locations in urban environments, and many of their projects are representations of how communities occupy and use urban public space. These projects offer an evocative portrait of the past and current life of the city, its residents, and visitors. Also in 2003, a mobile art project called [murmur][9] started an urban community storytelling movement, a concept that artist/designers Shawn Micallef, James Roussel, and Gabe Sawhney developed as students at the Canadian Film Centre's Media Lab. Signs depicting a large green ear with a telephone number inscribed on it started appearing on lampposts in Toronto's Kensington Market and Annex areas.

By calling the number on the sign with your cellphone, you can hear a short recording from someone with a story to tell about the house, back alley, market stall, synagogue, cinema, restaurant, club, or theater that you are standing outside. The stories are personal, and as diverse as the neighborhood itself. Green ear signs are popping up in international locations as far away as Edinburgh, Dublin, São Paulo, and San Jose – all designating local storytelling public art projects that have been co-developed with local communities and the [murmur] team.

The Haunting is an urban mobile game, developed by my colleagues David McIntosh and Michael Longford, with a team of student researchers and designers at OCAD and Concordia Universities. It was designed as a night-time adventure to be played in Parc Mont Royal, an urban park on the mountainside in Montreal. Developed as part of the Mobile Digital Commons Network (MDCN) in 2006, *The Haunting* is a ghost-capture game that uses Bluetooth beacons and the Mobile Experience Engine (MEE) software as well as GPS locative coordinates in the park. The aim of the experience is to capture ghosts with the phone, explore, and encounter other players; players keep track of each other's movements on the mountain via an interactive map that indicates their locations. An inventory counter reminds players how many ghosts they have left to capture, and players may gather together for a mobile séance to draw in more ghosts. Participants spend a thrilling night scouring the mountain park to encounter and capture the wayward spirits with their mobile devices.[10]

With the introduction of GPS and Bluetooth technologies as standard to mobile handsets, a world of possibilities opened up for short-range mobile to mobile networking, and less technologically encumbered experiences for mobile art participants. Some of these have been envisioned as participatory audio projects. *Alter Audio*, developed by my colleagues Paula Gardner and Geoffrey Shea as part of the Mobile Digital Commons Network (MDCN) in 2006, enhanced the urban environment with GPS-located vocal choirs and loops of ambient sound. Participants congregated in locations in Toronto and Montreal, and interacted together musically, or used gesture and Bluetooth proximity to collaborate with effects from the audio libraries on their mobile devices. In this way, even inexperienced participants could become mobile musicians, engaging in sound and music composition to create live musical collaborations.[11]

Mobile art and games can be used as a force for community building, critique, and change. *Park Walk* is a mobile public art project that I developed with my collaborator Bruce Hinds and my students in 2006. *Park Walk* is a social and environmental mapping project that delivers historical, cultural, and user-generated stories, all contributed by the local community, within the historic site of Grange Park near OCAD University in Toronto.[12] The project engages aspects of urban orientation and nature identification, local cultural activities, historical insight, and bio-regional mapping. The project lays a veil of information over the park, which is virtually without signage, creating a psycho-geography of shared public space.

Tactical Sound Garden (TSG) Toolkit, is an urban mobile audio intervention which considers the auditory culture of cities. It is a game based on open source software, developed by Mark Shepard, Professor of Architecture and Media Study at the University at Buffalo, SUNY. Shepard calls the *TSG Toolkit* a "parasitic technology" because it feeds on the propagation of Wi-Fi access points in urban environments as free, ready-made, and locative infrastructures for cultivating community sound gardens. *TSG* draws on the cultural paradigm of guerilla gardening to create participatory sound environments enabling urban dwellers living within wireless hot zones to install sound gardens for public use. This is done by planting sounds, which are mapped onto the GPS coordinates of the city, using a 3D audio engine more commonly used in gaming environments. Participants wearing headphones connected to their smart phone or Wi-Fi device drift though virtual sound gardens as they move around the city. In 2007, sound gardens were planted in Belo Horizonte (Brazil), San Diego, and Zurich.[13]

Rider Spoke is a mobile game for urban cyclists, designed by the British mobile art collective, Blast Theory.[14] The idea is to combine theater with cycling and mobile game play in a public urban environment. Participants bring their own bike, or borrow one from the artists. Cycling through the streets at night, equipped with a mobile attached to the handlebars, they find a hiding place to record a short message in response to a question posed, and then search for the hiding places of other participants' messages. *Rider Spoke* was created in

October 2007 for the Barbican neighborhood in London, and has been shown and played in Brighton, Athens, Budapest, Sydney, and Adelaide.

CherryBlossoms is a GPS-activated mobile art project, developed by Alyssa Wright at the MIT Media Lab, in Boston in 2007. Her project aims to build a more visceral sense of empathy for the victims of war. The project takes data from the locations of bombings in Baghdad and maps them with GPS hotspots to the streets of Boston, Massachusetts. Participants don a backpack outfitted with a small microcontroller and a GPS unit. Recent news of bombings in Iraq is downloaded to the unit every night, and the locations relative to the center of the city are superimposed via GPS coordination on the streets of Boston. If the wearer walks through a space in Boston that correlates to a site of violence in Baghdad, the backpack automatically detonates and releases a compressed air cloud of confetti. Each piece of confetti is inscribed with the name of a civilian who died in the war, and the circumstances of their death. Looking like a mixture of smoke, shrapnel, and the white blossoms of a cherry tree, the explosion completely engulfs the participant.[15] With *CherryBlossoms*, a mobile media project about human loss, the effect resonates far beyond the boundary of the original conflict.

In April 2008, a series of mobile public artworks were hidden in secret locations around the streets of Melbourne, Australia. Visitors used their mobile phones to find and interact with the *A-Lure* projects: digital projections, video and sound works, and a huge photographic light box. Clues were sent via mobile SMS (text messages) in a sort of public art scavenger hunt. Visionary Images (VI), a not-for-profit dedicated to engaging local youth who have experienced hardship, developed the project with over 140 young artists, teaching them skills to design and develop mobile media narratives and helping them to develop a sense of inclusion, power, and participation in their community.[16]

Conclusions

There are great challenges and barriers to the creation and deployment of mobile public art and games. These include the costs of developing, programming, and engineering mobile artworks for multiple platforms with variant compression codecs and programming languages. Although the iPhone application development programs have cracked open this space to innovation and whole variants of applications for specialized user groups are now surfacing, how much locative iPhone art are we seeing?

Can we expect to see practices for regulating consumption, limiting choice and innovation, and increasing corporate surveillance of our data, gain further momentum? There are looming issues regarding intellectual property and data sharing for mobile authors, service providers who distribute their data, and mobile participants who play and share, and we are all interested to see how these arguments will resolve.

Power struggles over ubiquitous file-sharing on mobile devices, with the intellectual property issues inherent in downloading, and with user-generated content, alongside the regulation of the wireless spectrum, all highlight the current dilemmas. The use and ubiquity of mobile devices in public space are so pervasive that issues of personal choice regarding always-on mobile technologies are only beginning to be addressed. This may be a conversation that it is too late to have, as children now grow up with mobile devices providing the ever-present soundtrack to daily life.

Frameworks that encourage community through collaborative artistic interchange are positive and unstoppable. These practices offer mobile artists, designers, audiences, and players fresh opportunities to develop exciting conversations and innovative experiences in the responsive mobile space.

Notes

1. Janet Cardiff works: http://www.cardiffmiller.com/artworks/walks/index.html

2. Masaki Fujihata works: http://medienkunstnetz.de/works/impressing-velocity/

3. Teri Rueb works: http://www.terirueb.net/trace/index.html

4. From Drew Hemment's writings on locative arts in the "TCM Locative Reader":

Locative Arts (2004). Reprinted in Leonardo vol. 39, No. 4 (August 2006).

5. RIXC Electronic Art & Media Lab: http://www.rixc.lv/

6. Locative Media Workshop at Karosta report: http://locative.x-i.net/report2.html

7. Rasa Smite & Raitis Smits works: http://www.parks.lv/home/E-LAB/archive/smiti.html

8. Frank Lantz: "Big Games and the porous border between the real and the mediated."

Vodafone Receiver. http://www.vodafone.com/flash/receiver/16/articles/indexinneroo.html

9. The [murmur] project: http://murmurtoronto.ca/

10. The Haunting: http://www.thehaunting.ca/intro.html

11. Alter Audio: http://mobilelab.ca/alteraudio/index.html

12. The Park Walk Project: http://mobilelab.ca/parkwalk/

13. Tactical Sound Garden: http://www.tacticalsoundgarden.net/

14. Rider Spoke: http://www.blasttheory.co.uk/bt/work_rider_spoke.html

15. Turbulence.org: http://www.turbulence.org/blog/archives/cat_tactical_media.html

16. The A-Lure Project: http://a-lure.org/art

EGO-TRAP
The Design and Implementation of a Digital Narrative

ANNE KAHR-HØJLAND

Viking Ship Museum, Roskilde

Digital narrative, augmented reality, mobile gameplay and serious gaming: the innovative exhibition concept titled *EGO-TRAP – you have no idea* goes by many names. In 2007, *EGO-TRAP* was launched as a cellphone-facilitated narrative pathway in the exhibition hall of a Danish science center,[1] marking the temporary high point of two years' design-based research.

This chapter opens with a description of *EGO-TRAP* as an educational design, then later presents some of the theoretical ideas lying behind the innovative exhibition concept, and concludes with some of the insights gained from the empirical studies in *EGO-TRAP*. In introduction, *EGO-TRAP* was designed with the aim of supporting pleasurable engagement as well as critical reflection in the exhibition. The latter was supposed to be triggered by pushing the boundaries of the conception of authority. As will be evident, this did not turn out exactly as expected.

EGO-TRAP: a digital pathway with two narrative layers

EGO-TRAP is probably best described as an individually adapted narrative pathway facilitated by visitors' personal cellphones. It works by means of WAP/GPRS technology and is characterized by three progressive levels and two narrative layers, referred to as the personal test and the meta-narrative, respectively.

Ben, a high school student from Denmark, describes his experience in *EGO-TRAP* as: *EGO-TRAP (...) to me it was almost like a journey, a trip (...) an ego-centric trip in a way, where you really got to know yourself (...) You think it is a trip, and then it turns out to be a trap.*

(Interview with Ben, Kahr-Højland 2010b, p. 9, translated from Danish)

Ben's feeling of "being trapped" hints that the experience in *EGO-TRAP* is pushing the boundaries of the conception of the museum as a genre as well as an authority. The following section describes the experience of *EGO-TRAP*.

First level: A personal test – and getting familiar with the procedure: On arrival the visitor signs up for *EGO-TRAP* in the entrance hall. The visit in the exhibition is initiated by a phone call, which the visitor will receive on his/her cellphone. A woman (e.g., the voice of a prerecorded actress which is distributed from a central server) will introduce herself and ask the visitor to play a simple computer game in which the visitor is to prevent a rat's escape from a maze. "The woman" will phone the visitor and explain that they are part of a personal test, in which she will guide the visitors from one hands-on exhibit[2] to another in the exhibition hall at the Experimentarium. "You say that human beings are smarter than animals," she states, "We would like to test that! First of all, I would like to find out how fast you are able to move in a wheelchair. Please go to the two wheelchairs in the exhibition hall." [3]

The woman guides the visitors to different exhibits in the exhibition hall by means of phone calls and maps, which are shown on the display of the visitors' phones. The maps display their current position and the next location. The participants are sent to different exhibits in order to create individual pathways and avoid queuing up in the exhibition hall. Each exhibit represents a post in the personal test; before interacting with the exhibit, visitors are asked to make a prediction of their own results by typing it on the keypad of their phones. *Wheelchair Energy* represents one example of an interactive exhibit in *EGO-TRAP*. At this exhibit you are supposed to move in the wheelchair as fast as possible. While you move the wheelchair, lemonade drips into a glass. The lemonade corresponds to the energy the visitor uses while moving the chair. When the visitors finish, they can regain the energy lost by drinking the lemonade.

After each exhibit in *EGO-TRAP*, the instructional voice will phone visitors and evaluate their performances. The voice might say: "You tend to underestimate your own physical strength. Let's see what it's like when it comes to recognizing colors..."

The personal test represents the primary narrative layer in *EGO-TRAP*. After four exhibits the visitors will receive a personal profile constructed from their performances in level 1.

Based on this profile, the instructional voice will match the visitor with another person in *EGO-TRAP*. If the visitors agree to be matched with a partner, they will move on to the next level of *EGO-TRAP*, which is the level of cooperation.

The second level: about cooperation – and a growing suspicion that the test is not what it seems to be: In level 2 the visitors are still navigating within the framework of the personal test. Just as in level 1, the visitors are guided to different exhibits, they type their predictions, and the instructional voice subsequently assesses their performances. But unlike level 1, the scope of the personal test in level 2 is to uncover the visitor's ability to cooperate with his or her matched partner. In level 2 the visitors, among other things, are asked to whisper secret messages to each other using two satellite dishes and cooperate to score points in a game called Play with the powers of the air.[4]

Level 2 is also the level that arouses suspicion, as the visitors will realize that their experience in *EGO-TRAP* is not the personal test it seems to be. At a certain point the participants receive a phone call from another instructional voice, this time that of a man, who explains that he has hacked into the system. "The hacker" throws suspicion on the woman, the original instructional voice: "Why is she asking you all these questions? I think she may be playing games with you," he states. He urges the visitors to cooperate with him in bringing the woman's true agenda to light. The hacker's entrance on the scene indicates the presence of another narrative layer in *EGO-TRAP*: namely, the narrative layer referred to as the meta-narrative.

Third level: Confrontation – and the disclosure of who's really behind the test: The visitors have to decide whether they trust the hacker or the original instructive voice. If they believe the woman, she will be very eager to make them finish the game and leave *EGO-TRAP*. If they believe the man, he will guide them to what he assumes is the woman's secret room. The room is situated underneath the central staircase in the entrance hall of the Experimentarium, right next to the desk where the visitors signed up for *EGO-TRAP* and played the *Ratrace* game. The room appears to be locked by a secret code, but the hacker breaks the code and allows the visitors to enter. When the visitors enter the small dark room, the door slams behind them, and they are confronted with an animated rat, which speaks with the voice of the woman. It turns out that the woman is in fact a rat who has mutated. She and her fellow animals have taken control of the science lab and are now investigating humans. As a matter of fact, the roles have been switched: the participants have been in the roles of testing animals for the past 1.5 hours in *EGO-TRAP*. The rat is angry that the visitors found her secret place and she declares that they will be trapped in her cage unless they are able to conquer her in *Ratrace* – a 3D version of the computer game that they played initially in *EGO-TRAP*. Now the visitors are in the role of the rat trying to escape the maze, while the rat/woman is the one closing the doors. When the visitors conquer the rat (at some point they will, as the game was created so that the visitors are predetermined to win), the door will open and they are free to go. This is where the experience in *EGO-TRAP* ends.

The technique behind EGO-TRAP

The technological development of *EGO-TRAP* has been carried out in cooperation with a Danish firm, Unwire. The visitors' phones must be able to run WAP/GPRS, which is a free service; however, you pay for the amount of information you download. The experience in *EGO-TRAP* lasts for about 1.5 hours and during that period of time you download information to a total cost of approximately 1 USD.

The communication in *EGO-TRAP* runs in a WAP/GPRS session and is characterized by being a two-way dialogue. A central server distributes the phone calls from the woman and the hacker, as well as the guiding maps on the display; the visitor responds to the system by means of the keypad on the phone. During the *EGO-TRAP* experience, the participants are still able to receive phone calls from other people. They can take a break, visit other parts of the exhibition for as long as they want, and still return to the place where they stopped in *EGO-TRAP*.

Design Based Research as a methodological approach: Developed as part of a Danish research program, DREAM (Danish Research Center on Education and Advanced Media Materials), *EGO-TRAP* was designed as an innovative learning resource, using Design Based Research (DBR) as the overall methodological framework (Baumgartner, et al, 2003; Barab & Squire, 2004). The basic idea of DBR is to bridge the gap between educational theory and practice. Working within the framework of DBR implies that you as a researcher make changes – for example, implementing new educational designs in an existing learning context (in this case the Experimentarium). The educational design is planned on the basis of theoretical arguments developed by the researcher, whereas the development and implementation of the design is carried out in close collaboration with practitioners in the learning context. The design process of *EGO-TRAP* was carried out as an iterative design process counting a total of four iterations. For further details on the methods and design process see Kahr-Højland, 2010b, or contact the author).

Why was EGO-TRAP made?

EGO-TRAP was originally planned as a learning remedy at a science center, with students aged 14-17 from high schools as the primary target group. *EGO-TRAP* was developed with the aim of supporting pleasurable engagement and critical reflection. As argued for by Kahr-Højland, 2010b, pleasurable engagement and critical reflection are to be considered important constituents of a modern conception of scientific literacy (Kahr-Højland, 2010b).

The classical science center and the need of scaffolding: The design process of *EGO-TRAP* started with the pinpointing of a problem: the lack of structure in the exhibition hall at the Experimentarium. The Experimentarium is considered a "classical science center" i.e., a science center that builds on the tradition of Frank Oppenheimer (Oppenheimer, 1968). In his *Rationale for a Science Museum*, Oppenheimer claimed that visitors at science centers were to walk into a "forest of phenomena" (that is, hands-on exhibits) where they should be free to interact with exactly the kind of phenomena that caught their interest. Oppenheimer's rationale foreshadowed the opening of The Exploratorium in San Francisco (1969). Since then, thousands of science centers influenced by the ideas of Oppenheimer have seen the light all over the world. To visitors, the "classical science center" exhibitions appear as an explorative milieu with no tangible structure (Bradburne,

2001; Kahr-Højland, 2010b); these kinds of exhibitions have proven to cause a lot of "random button pressing." The average time spent on hands-on exhibits at science centers have been measured as less than 30 seconds (Walker, 2008, Paris, 1997). The design process of *EGO-TRAP* therefore started with the assumption that if classical science centers like the Experimentarium have a learning goal which is higher than that of an amusement park, there is a need for scaffolding[5] – i.e., a need to offer directional guidance – to a higher extent than in the Oppenheimer tradition (Kahr-Højland, 2010b, Kahr-Højland, 2010a). The *EGO-TRAP* educational design experiment was based on the idea that scaffolding the visitor, by adding a digital narrative pathway to the existing exhibition, might solve the problem of random button pressing.

The narrative as the primary remedy of scaffolding: The narrative was chosen as the overall means of scaffolding in the educational design. The choice of the narrative as a scaffolding remedy at the Experimentarium was primarily grounded in results from cognitive science (represented by, among others, Jean Mandler, 1984, and Jerome Bruner, 1986; 1990) and sociolinguistics (represented by William Labov, 1972). The insights gained from these theoretical fields point to the fact that narrative possesses some intrinsic qualities due to the structure. Results from cognitive science have thus shown that narrative and human comprehension are indeed closely related. We use narrative structure in order to understand ourselves and our surroundings (Mandler, 1984;

Bruner, 1986; 1990). In his sociolinguistic investigations, Labov has pointed to the fact that the narrative has an integrated dimension of evaluation. Labov characterizes the element of evaluation as "perhaps the most important element in addition to the basic narrative clause" (Labov, 1972, p. 359). Due to this built-in element of evaluation, the narrative establishes a distance between the audience and the narrative. And by that, Labov argues, the narrative in itself calls attention to the fact that it is indeed a narrative – which again supports meta-reflective processes within the audience (Labov, 1972).

The intrinsic qualities of the narrative: Mandler (1984), Bruner (1986; 1990) and Labov (1972) show some of the intrinsic qualities of the narrative:

- The narrative organizes information in a meaningful way as it provides a recognizable context and by that, supports human comprehension. (Mandler, 1984, Bruner, 1990)
- The narrative represents a well-known framework which is decoded easily by a broad target group. (Mandler, 1984; Bruner, 1990)
- The narrative holds the possibility of containing several systems of meaning, running in parallel: this means that the narrative can function on an elementary level of action as well as on an additional level of affection (designated by Bruner as the landscape of action and the landscape of consciousness, respectively). This makes it possible

to engage people on several levels (cognitively and affectively) and makes it possible to challenge a broadly defined target group. (Bruner, 1986)

- The narrative scaffolds reflection by encouraging different interpretations rather than calling for clear-cut answers. (Bruner, 1990)
- The narrative calls attention to itself as being a narrative, and by that it scaffolds meta-reflection. (Labov, 1972)

The potential of the narrative: The intrinsic qualities are all ascribed to the structure of the narrative. Additionally, as argued for in Kahr-Højland (2010b), the narrative possesses a potential for supporting both what Mihalyi Csikszentmihalyi calls flow experiences (Csikszentmihalyi, 1990) as well as social learning processes (Wenger, 2003).

Flow: Flow may be best described as the positive feelings our body gives us in reward when we learn something new. According to Csikszentmihalyi and Hermanson, flow is:

> *A state of mind that is spontaneous, almost automatic, like the flow of a strong current (...). When goals are clear, feedback is unambiguous, challenges and skills are well matched, then all of one's mind and body becomes completely involved in the activity (Csikszentmihalyi and Hermanson, 1995, p. 70)*

The achievement of (ideal) flow is about adequate challenge. Because visitors have individual capabilities, flow is all about framing experiences that are flexible and user sensitive. In that way it is possible to adequately challenge a broad target group and thereby make them engage pleasurably in the exhibition.

By experimenting with the balance between the known and unknown, the narrative may support flow experiences; due to its well-known form ("Once upon a time ..."), the narrative invites a broad target group and subsequently challenges the audience, either through the development of an unknown content or by means of alternative narrative structures. Thus, the narrative is characterized by representing something well known and yet something unknown (we don't know the story before the narrative has been carried out).

Social learning processes: By creating a suitable balance between the known and unknown, the narrative possesses a potential for scaffolding flow experiences. Also, the narrative has potential for supporting social learning processes by providing a framework that calls for social participation. Etienne Wenger has shown how knowledge is produced from a constantly ongoing negotiation – verbally and non-verbally – between individuals and their surroundings. The negotiation of meaning takes place in different kinds of communities of practice (Wenger, 2003).

The narrative possesses the potential for supporting social learning processes by imitating communities of practices (Kahr-Højland, 2010b). To realize this potential, a special kind of narrative is required. The narrative in *EGO-TRAP* was thus planned with the aim of using the intrinsic qualities of the narrative, as well as realizing the potential of the narrative by imitating the establishment of communities of practice (Kahr-Højland, 2010b).

The presence of the narrative in EGO-TRAP: Jerome Bruner has established the following minimum demands of what constitutes a narrative: a narrative is characterized by having an actor who acts in order to reach a goal in a recognizable environment. The actor uses different kinds of means in order to obtain his goal. Often there will be a problem, which propels the action from an initial position to a final position (Bruner, 1996, p. 161).

With this definition of a narrative in mind, *EGO-TRAP* is characterized by two narrative layers, referred to as the personal test and the meta-narrative.

The personal test: In the personal test the visitor takes up the role of the actor, who acts in order to reach a goal. Framed by the apparent narrative framework of a test, the goal is to gain insight into your own abilities (manifested in a personal profile). The exhibition constitutes the environment, while the exhibits and the woman (original instructive voice) represent the tools used by the actor in order to reach the goal. The initial position is represented by the participant's insufficient knowledge about himself, while the final position appears when the participant receives his personal profile.

The purpose of the test narrative was to create a narrative framework that would invite the young target group to engage pleasurably by offering them a role as a subject, with which they could easily identify. This was based on previous studies of the young target group as well as focus group interviews carried out initially in the design process (Buckingham & Scanlon, 2003, Quistgaard, 2006, Kahr-Højland, 2010b). Also, the idea was to create a narrative framework that emerged naturally from the already existing exhibition at the Experimentarium. The test narrative made it possible to involve predictions and evaluations as a natural part of the design, which again made it possible to carry out the narrative's potential for supporting flow experiences.

The meta-narrative: In the meta-narrative the visitor is still the actor, but the goal has changed: now it is all about gaining insight into the underlying objectives of the setup (the disclosure of the woman/rat). The means to gain insight are represented by the partner, the hacker, and the cellphone (which can be used to phone others, asking for advice). In the initial position, the visitor is unaware that the woman is a rat, while in the final position, the visitor is confronted with the woman/rat in the rat's room.

Ideally, the meta-narrative was thought to establish a narrative layer about conspiracy theory, which would prompt the visitors to take a step backward and reflect critically on the setup of *EGO-TRAP*. The hacker's entry on the scene marks an unexpected turn in *EGO-TRAP*. By experimenting with the content and form of the narrative, the idea was one of challenging the visitors in order to carry out the narrative's potential for supporting flow experiences. Similarly, working with a partner as a team in *EGO-TRAP* was an attempt to carry out the narrative's potential for supporting social learning processes. The cooperation in *EGO-TRAP* happens within the framework of the personal test. However, at the same time, the partnership was thought to simulate a community of practice. The ideal goal in this case was to urge the students to discuss a shared repertoire of exhibits and, in that way, encourage critical reflection in relation to the experience within the exhibition.

The meta-narrative was thought to engage the students and to support critical reflection by challenging the concept of the museum as an authority. Ideally, *EGO-TRAP* was thought to support reflection on the foundation for knowledge. The revelation that the woman is a rat represents an element of surprise, which was thought to challenge the idea of the museum as an authority. The idea was to call attention to the fact that information is not considered an absolute truth just because it's presented by a science center or a museum. Thus the intention for the metaphor of conspiracy theory in *EGO-TRAP* was to identify the fact that there is always someone behind the information you receive; the sender, who presents the information, has always inevitably interpreted the information we receive . By pointing out the fact that the sender is not whom she pretends to be, the meta-narrative in *EGO-TRAP* was thought to challenge young people's basic ideals of knowledge.

Empirical studies

Following the guidelines of Design Based Research, *EGO-TRAP* was implemented in the exhibition hall at the Experimentarium through a design process with a total of four iterations (Kahr-Højland, 2010b). The adjustments of the educational design were based on empirical studies of young people interacting in the exhibition. The empirical data collections and the subsequent data analyses were planned as a qualitative study (Flick, 2006; Kvale, 1997). The empirical studies in *EGO-TRAP* were designed so as to gain insight into young people's experiences and reflections related to their interactions with the exhibits at the science center

In total, more than 300 students tested *EGO-TRAP*. These students came from both primary and secondary schools. The students tested *EGO-TRAP* in different stages of the iterative design process as every test round resulted in adjustments in the educational design (Kahr-Højland, 2010b). Many of the students (about 200) were observed in the exhibition in order to make adjustments to the educational design. Some of these (about 50) participated in focus group interviews at the Experimentarium immediately after their visit to *EGO-TRAP*.

The primary group of subjects consisted of three high school classes, selected from different areas of Denmark. From each class two subjects, one male and one female, were selected for video recordings and subsequent semi-structured interviews, each lasting for about one hour. These six were the key informants of the study. The selection of the key subjects were based on the criteria that they should be average science students and that they would not be too disturbed or anxious by the fact that they were supposed to be video recorded. The interviews were carried out one to two weeks after their visit to *EGO-TRAP*. The interviews were planned as semi-structured interviews, in line with the suggestions by Steinar Kvale (Kvale, 1997).

The data analyses were based on transcripts of approximately ten hours' video recordings and eight hours' interviews with the key informants. The video transcripts were completed with the aim of covering as many semiotic levels as possible, in order to prepare an analysis that was as open and explorative as possible. The transcriptions describe among other things, the subjects' mimics and gestures; their movements in relation to each other, the exhibits, and the cellphones; as well as their dialogue during the experience in *EGO-TRAP*. Based on the theoretical framework of the project, the analysis was carried out with special regard to investigate how the narrative layers work in relation to the engagement of individual students and the possibility of achieving flow – as well as their development of critical reflection by means of social learning processes. Following the suggestions of Brigitte Jordan and Austin Henderson, the data was coded following both a vertical and a horizontal structure (Jordan & Henderson, 1995).

Insights gained from the empirical studies

Seen from the narrow perspective of an exhibition designer, the design of *EGO-TRAP* may be described as a success. The young visitors are happily engaged users who navigate comfortably and securely in the exhibition with their cellphones. *EGO-TRAP* scaffolds young people in the exhibition so as to avoid the random button pressing that often characterizes a visit to a science center. Insights from the data analyses also indicate that the personal test supports experiences of flow. The young people's engagement is primarily linked to the narrative layer that concerns the personal test.

The personal test frames the exhibits in a meaningful way: A narrative framework inevitably offers a certain role or subject position for the person who interacts with it (Goffman, 1959; Bruner, 1996; Gergen, 2002). The video recordings of one of the key subjects in *EGO-TRAP*, Ben, suggest that the position offered by the narrative framework of the personal test is one that Ben can easily relate to. As an example, when Ben arrives at *Wheelchair Energy* a group of small girls, aged about 7, are playing at the wheelchairs. The video recordings show how the 17-year-old Ben sits down alongside the small children. In order to achieve a good result at the exhibit, Ben gives everything he has: he wheels the chair until he reaches the pain threshold in his arms, showing signs of pain when he finally reaches the goal. Seen from the outside, the situation with Ben – a tough high school guy – resembles the situation of a young man sitting down in the sandpit surrounded by toddlers. But apparently, Ben does not experience the situation in this way. He is deeply engaged in fulfilling his role in the personal test. Ben himself describes the sequence at the wheelchairs very emotionally: "at that time it was like it (the exhibit) just wanted to make me bite the dust" (Kahr-Højland, 2010b, p. 71). Ben identifies with the role of a test person in a way that makes it meaningful and relevant to him, even in an exhibit that is practically inundated with toddlers.

The personal test supports flow by balancing known and unknown: The participants' behavior and descriptions of the *EGO-TRAP* experience indicate that the guidance by the original instructive voice provides clear goals and unambiguous feedback by means of a well-known structure (prediction – performance – evaluation). At the same time, the personal test is characterized by having unknown content. The young people have a serious approach to the exhibits, which they often describe as providers of (unknown) information about themselves. As one of the informants, Anna, puts it:

> *Entering your result and getting feedback was really funny, I also found it very funny that you did not know what was next to come, I mean, as compared to receiving working sheets which tell you to do like this and then to do that (...) in a sense, you were totally unprepared of what was next to come, what you should test and what you were supposed to answer, and that, I think, was very amusing. (Interview with Anna, Kahr-Højland, 2010b, p. 70 translated from the Danish)*

Social learning processes: The data collections also indicate that *EGO-TRAP* realizes the potential of the narrative when it comes to supporting social learning processes. As will be evident in the following paragraph, this does not happen – as was expected – with verbal negotiations of meaning. Instead, the personal test tends to scaffold non-verbal negotiations, as the young people negotiate identity through their performances in *EGO-TRAP*. The young people's negotiations of identity through the personal test indicate that individual and social learning processes are in fact two sides of the same coin.

The engagement in the personal test happens at the expense of the meta-narrative: The young people's strong commitment to the personal test happens at the expense of their engagement in the meta-narrative, which was planned to support critical thinking through social learning processes. The video recordings of Ben and his partner, Mille, at the exhibit *Reaction and Overview* serve as an example that the students do not demonstrate the same level of commitment to the meta-narrative as to the personal test. *Reaction and Overview* consists of a wall on which flashing lights appear in random order. Each time you press a button with a flashing light it will be switched off and a new light will begin to flash. The task consists of pushing as many flashing buttons as possible within 60 seconds. This exhibit appears in the second level of *EGO-TRAP*. When arriving at *Reaction and Overview*, Ben and Mille have been through different kinds of tests while, concurrently, the meta-narrative has developed and is now about to reach its climax. The hacker keeps calling with circumstantial evidence that the woman has a hidden agenda. He urges them to phone other visitors in *EGO-TRAP* in order to figure out the truth about the woman. But Ben and Mille are much more occupied with getting a good result in the test. They are performing to each other as well as to their surrounding classmates, and the only evidence we get that the hacker is calling them and urging them to contact other classmates, is Ben's comment: "I haven't (...) it came up with all these phone numbers, I haven't phoned anybody." No one answers him and they never discuss it again. As it seems, Ben and Mille continue in their roles as test persons, which they have adopted since they were first introduced to the personal test. At this point, despite the theme of conspiracy introduced by the hacker, they are still in their roles as test participants, occupied with their progress in the test.

Performing a third narrative layer: The young people do engage in the meta-narrative in the sense that they express having fun and physically act on what Bruner refers to as the landscape of action of the meta-narrative (c.f. the intrinsic qualities of the narrative p. 142); they actually run to the rat's room at the end of the sequence. But their engagement in the meta-narrative never reaches the high level of engagement, evident in the part of the experience related to the personal test. When Ben and Mille perform at *Reaction and Overview*, they do it within the framework of the personal test. Apparently, they feel comfortable with the subject positions offered by the narrative framework of the test and as a consequence, they do not negotiate meaning through critical dialogue about the hacker and the woman. Over ten hours of video and mp3 recordings from the students' interactions in *EGO-TRAP* reveal the fact that in *EGO-TRAP*, verbal communication is reduced to a minimum. This means that the critical discourse and negotiation of meaning that was thought to be prompted by the meta-narrative does not emerge. This does not mean that the social learning processes are absent in *EGO-TRAP*; it rather means that the social learning processes are happening in unexpected ways.

The situation at *Reaction and Overview* is one of many situations in *EGO-TRAP* indicating that the students are in fact performing to each other in *EGO-TRAP* (Kahr-Højland, 2010b). When Ben performs his best, it is not just to get a good result at the exhibit. The subsequent interview with Ben reveals that he is very much aware of the fact that the classmates are watching him, judging his performance. Thus the result at a given exhibit is not just a personal matter: every single result represents a piece in the jigsaw puzzle constituting the identity of the student *(Am I strong? Do I have musical abilities? Am I good at estimating my own abilities?)*. Every single piece of identity is obvious, available to everybody, and is immediately thrown out in a larger jigsaw puzzle constituted by the social context – or community of practice – formed by the high school class. This social gameplay represents a third narrative layer in *EGO-TRAP*. The data analyses from *EGO-TRAP* thus point to the fact that the students are negotiating identity through non-verbal performances at the individual exhibits. This non-verbal negotiation of identity is framed and determined by the social gameplay and is identified as the third narrative layer in *EGO-TRAP* (Kahr-Højland, 2010b).

Why did they not engage in the meta-narrative to the same extent?

One of the reasons the students do not commit as strongly to the meta-narrative as to the personal test may be ascribed to a built-in problem in the design of *EGO-TRAP* – namely, the apparent clash of genres between the personal test and the meta-narrative.

A clash of genres: The students visit the Experimentarium as part of a school trip. On arrival they are introduced to a test, which is basically positivistic. Altogether, the setup is not far from a formal school setting. Students are familiar with this formal framework and indicate that it makes them feel comfortable. Ella, a girl aged 17, explains that she:

> *(...) experienced (...) great confusion, (but) the woman called me and she was very good at, like, calming things down, and also, like, enabling us to relax in the situation, because she kept calling us, telling us what to do. (Interview with Ella, Kahr-Højland, 2010b, p.69, translated from Danish)*

The students know exactly what is expected from them (that is, to perform in a test) and what they themselves expect from the experience (that is, to get a clear-cut answer from the test).

When the hacker enters the scene, it marks a dramaturgical turn – and also, in fact, a radical change of genre as the hacker/rat story resembles the movies and computer games watched and played by the students in their spare time. The students express how the emergence of the hacker in the story confuses them: Ella states that "it wasn't exactly what I was expecting (...) it was actually pretty surprising" (Kahr-Højland, 2010b, p. 75, translated from Danish) while Finn, aged 16, explains that "I didn't understand the story at all at that time (...) All of a sudden the whole thing got a little too advanced." (Kahr-Højland, 2010b, p. 75, translated from Danish)

Playing with the conception of authority: So, the hacker tends to stagger the balance between known and unknown, which was established by the personal test. The idea was one of urging the students to take a step back and look critically at the setup, using the competencies they normally use when solving problems in computer games. But this critical, investigative perspective collides with the positivistic framework of the personal test and thus leaves the students in a stage of confusion and frustration.

The surprising clash of genres in *EGO-TRAP* does not account alone for the students' frustration and confusion. Let's play with the idea that *EGO-TRAP* had been implemented as an entertainment feature in a shopping mall. The point being, that the involvement of the hacker does not represent something innovative or surprising per se; in many ways the hacker story resembles the kind of movies preferred by the young target group. A focus group of students from high schools unanimously stated that among their favorite films were movies having an element of either conspiracy theory, or a surprising revelation implying a fundamentally changed perspective on the storyline. Movies like *The Matrix*, *The Sixth Sense*, and *A Beautiful Mind* were all mentioned as examples of this kind of movie (Kahr-Højland, 2010b). It is the fact that the hacker-story is implemented at a science center that is innovative and astounding. The rat/hacker story does not just represent a clash of genres as compared to the personal test, it also stirs up the concept of authority as it relates to the science center (as compared to, for example, a shopping mall).

When the students arrive at the science center, together with their classmates, they submit to the (voice of the) woman. The woman represents the science center as an institution and the students express that they regard her as an authority, which they follow blindly. They also express that they feel comfortable about this, as it resembles a well-known situation at school (Kahr-Højland, 2010b).

The science center as an authority – and as a Paradigm II museum: As science centers are traditionally associated with high professional expertise, the students look upon the woman as a figure of great authority. Science centers have been classified as so-called Paradigm II or Mode 2 museums (Pedretti, 2004; Nowotny et al , 2003, Kahr-Højland, 2007). Science centers were originally introduced as an antithesis of the traditional museum, also referred to as Paradigm I museums, in the 1960s. In traditional museums the emphasis is predominantly on the information provided rather than on the visitor. The traditional museum was built on the idea of enlightenment and the collection is often identical with the exhibition. Emerging from the idea of collecting information in order to make it available to the public, the educational point of view in the traditional museum is one of public outreach (Hooper-Greenhill 1999; Parry 2007).

The model below (Fig. 1), suggested by Kahr-Højland in 2007, provides an overview of three museum paradigms of which the third is yet to come.

The opening of the first science centers in the late 1960s marked a veritable paradigm shift within museum

	Traditional museum (I)	Interactive Museum (II)	Augmented Reality Museum (III)
Primary focus	Information	Audience/receiver	Information + audience
Media	Display case (+ boards, movies, DVD recordings etc)	Hands-on exhibits	Interactive exhibits + narrative structure + mobile technologies
Learning approach	Positivist	Constructivist	Social constructivist

Figure 1: Three different museum paradigms – an overview. (Kahr-Højland, 2007, p. 136)

communication. Based on the idea of Frank Oppenheimer, the science centers were planned to make the visitor construct her own knowledge by manipulating things in hands-on exhibits. Gone were the signs of *don't touch* and *keep silent* as well as the structured, guided tours. The classical science center with the element of free play at the core was born, and the success was so massive that the idea of interaction spread to traditional museums (Bradburne, 2001). Due to its emphasis on free play and interaction, the classical science center thus marked a change – a paradigm shift – within museum communication, a paradigm stating that you are not only allowed to touch, you are actually supposed to touch and by that construct knowledge about a phenomenon.

The learning approach thus shifted from a somewhat positivistic approach to knowledge in the traditional museum to a constructivist approach in the science center/Paradigm II museum. But even if the approaches to achieve knowledge are disparate in the two museum paradigms, they both build on a basic ideal of knowledge, implying that "we" – the educators – know what "you" (the public) should know. Science centers and traditional museums are therefore both associated with a high degree of authority, as both kinds of museums present themselves as strongholds of knowledge.

The third paradigm museum – facilitating critical reflection: Today, we are experiencing a fundamental change in our cognitive conditions: knowledge is no longer considered as universal truth, but rather something that is open to debate (Nowotnyet al, 200). As Barnett (2004) has argued, the students of today have to cope with the fact that they are "knowing-in-and-with-uncertainty." It has therefore been emphasized from different theorists, as well as practitioners working in the field of museum learning, that there is a need for a third paradigm museum, which facilitates open debate and critical reflection to a higher extent than is the case of the science centers (Henriksen & Frøyland, 2000, Pedretti, 2004, Bandelli, 2005, Quistgaard & Kahr-Højland, 2010b). A third paradigm museum that aims at supporting open debate and critical reflection by means of interactivity and flexible narrative structures facilitated by mobile digital technologies (Pedretti, 2004; Bandelli, 2005; Kahr-Højland, 2007, Quistgaard & Kahr-Højland, 2010b).

The third paradigm differs decisively from the previous two paradigms due to its prompting of social meetings and peer-to-peer communication. Traditionally, the museum experience has been predominantly individual, as visitors in Paradigm I museums have been expected to silently study the items displayed on their own. On certain occasions the introduction of digital technologies has even emphasized the individualized aspect of a museum visit. Audio guides, as an example, often tend to hinder conversations between visitors. Similarly, despite the explicit call for playing and talking, Paradigm II museums (i.e., science centers) are ultimately focusing on the personal interaction between the individual and the exhibit (Kahr-Højland, 2007).

EGO-TRAP as an attempt at realizing the third paradigm museum by challenging the science center as an authority: We are seeing the contours of a third paradigm museum emerging, as pointed out in Kahr-Højland, 2007 and Quistgaard and Kahr-Højland, 2010b. A paradigm reflecting the fact that "we" (the educators) no longer know exactly what "you" (the visitors) should learn. The audience is to realize that knowledge is open for debate and thus should be scaffolded in questioning the basic premise for presenting information within a museum. As argued by Quistgaard and Kahr-Højland (2010b), mobile digital technologies hold great potential as facilitators of personalized, as well as socially mediated, learning processes within science centers and museums. The potential of mobile digital technologies as key players is primarily grounded in four qualities of these technologies, identified as mobility, flexibility, communicative property, and media transparency (Quistgaard & Kahr-Højland, 2010b).

EGO-TRAP may be seen as an attempt to realize the third paradigm museum, as it does not aim to present definite truths. Rather, it aims at facilitating processes of inquiry and critical reflection. It does so by providing a semi-structured experience in an explorative milieu, which should scaffold pleasurable engagement (flow), social learning processes, and critical reflection. The latter was thought to happen through verbal negotiations of meaning between the team partners in *EGO-TRAP* – initiated by the surprising challenge of authority.

EGO-TRAP challenges the conception of authority: The meta-narrative in *EGO-TRAP* does not just mean a change in genre as compared to the personal test. The idea of exposing that the "museum guide" (i.e., the woman) whom visitors thought would reveal scientific truths about themselves, is not who she pretends to be, was to give the visitors a great surprise – to give them a shock that would change their view on the communication situation at the science center and make them reflect critically. The disclosure of the woman as a rat was intended to urge the participants to question the basic premises for communication at the science center. But the scaffolding of this critical questioning of the basic premises for knowledge could have been more tangible than was the case in *EGO-TRAP*.

Some of the students in *EGO-TRAP* do actually question the premises for information provided by the exhibits. As an example, Ella, aged 17, gets very upset because one of the exhibits concludes that she is prejudiced. Ella does not understand the premises for the conclusion – but even though she is disturbed and questions the validity of the conclusion, and even though she actually states that the experience in

EGO-TRAP has taught her to "reflect critically" and to "always be aware of who the sender is," she keeps talking about the result as "scientific proof." She states that the experience in *EGO-TRAP* has scientifically proven what kind of person she is (Kahr-Højland, 2010b). So, even when the exhibit result goes against everything she knows about herself, she ends up trusting it because she credits the woman/science center as a greater authority.

The third way of authority: According to the Danish educational researcher, Lars Qvortrup, authority is basically about taking care of someone, much like a parent who takes care of a child by setting up rules for bedtime. In the past, authority was associated with uniforms and institutions. Then, after the anti-authoritarian period, authority was defined psychologically as personal authority. Today, Qvortrup states, we are approaching the third way of authority, where authority is related to competencies, for example your educational competency or your personal competency (e.g., as a doctor or as a parent) (Qvortrup, 2010). Authority may be challenged, for example, in a classroom where the teacher may be confronted with viewpoints that are opposite to his or her own. By challenging authority, it may be either weakened or strengthened. Authority in a modern classroom is all about facilitating and steering discussions where everybody is allowed to express his or her viewpoints. This is not the same as creating a situation where anything goes. The teacher uses professional expertise – i.e., his or her authority – when facilitating the discussion.

The idea of the meta-narrative was to challenge the idea of the museum/science center as an authority. But challenging the concept of authority is not the same as weakening your responsibility as an authority. Actually, it means that you have to strengthen your responsibility as an authority – for example, by facilitating open-ended debates and by that, scaffolding critical reflection. As an example in *EGO-TRAP*, the science center could have assumed its responsibility as an authority for the students by facilitating an open debate about *EGO-TRAP* in direct continuation of the experience. This could have been put into practice, either virtually in a chat room or physically in a discussion group, where students as well as teachers could contribute to the debate facilitated by staff members from the science center. An even simpler solution could have been carried out in the form of audio-recorded video clips, where visitors could express their thoughts and feelings when exiting the rat's room. By publishing the video clips immediately on an electronic wall, the comments of the participants would contribute to an ongoing debate about *EGO-TRAP*.

Conclusions

In some ways, *EGO-TRAP* succeeds as an exhibition concept foreshadowing a third paradigm museum. The use of a flexible narrative structure facilitated by cellphones does indeed scaffold the visitors in a way that engages them pleasurably within the exhibition without compromising the explorative approach. The students report on the experience in *EGO-TRAP* with great excitement. They spent more than 1.5 hours in the exhibition, trying the exhibits again and again. So, *EGO-TRAP* succeeds in solving the problem of random button pressing identified as a common problem at classical science centers.

However, when it comes to the ideal goal of scaffolding critical reflection by challenging the conception of authority, *EGO-TRAP* could have emphasized the science center's role as an authority to a greater extent. The idea was to support critical reflection and to challenge the existing ideal of knowledge. This implies a huge responsibility on the museum to hold visitors and guide them in their process of critical reflection and their questioning of the ideal of knowledge. By scaffolding open-ended debates in more tangible ways, *EGO-TRAP* might have supported the questioning of the premises for knowledge more effectively.

The museum may play with genres and challenge the concept of how to conceive and relate to knowledge, but the shaking up of authority within the museum does not mean that the museum should disclaim responsibility altogether. In fact, museums should take on their responsibility as authorities to an even greater extent by developing and supporting scaffolding remedies by which visitors can better understand and create information.

Notes

1. EGO-TRAP was implemented in the main exhibition at the Experimentarium in Copenhagen, Denmark, in 2007. It will be open for the public until 2011 (http://www.experimentarium.dk/EGO-TRAP).

2. The term hands-on exhibits is used throughout this article covering the kind of exhibits used in science center exhibitions. It is an ongoing discussions at science centers worldwide whether to use the term hands-on exhibits or interactive exhibits. Many researchers working with science center learning argue for the use of the term interactive exhibits (c.f. Rennie & MacClafferty, 1996) while most science centers tend to stick with the original term hands-on exhibits (for example Tullahoma Hands On Science Center, Tennessee, Natural History Museum, Los Angeles, Science Museum, London).

3. The visitors are sent to different exhibits in order to avoid queuing up on the exhibition. Six exhibits (of which each visitor visits four) form level 1 in EGO-TRAP.

4. At the Satellite Dishes, the visitors are supposed to whisper messages to each other over distance by means two satellite dishes: if the visitors speak directly into the focus of the dishes, they are able to hear each other whispering over a distance of several meters. At Play with the Powers of the Air, the visitors are to play ball with a pair of blowing hoses. Each visitor has to balance a ball on an air stream and attempt to pass the ball through a ring by adjusting air velocity.

5. Since the term scaffolding was introduced by Jerome Bruner in the 1950s, it has been used in a widespread set of contexts and with various meanings. Building on Vygotsky's theory about the zone of proximal development, Bruner et. al. describes a scaffolding as the framework (scaffold) a teacher establishes for a learner in order to help the learner to concentrate upon those elements that are within his range of competence (what Vygotsky call zone of development), (Wood, Bruner, Ross, 1976, p. 90). In this context the term scaffolding is used in the broad sense of providing "instructional guidance" for a learner. (Kahr-Højland, 2010b)

References

Bandelli, A. 2005. Fostering deliberative democracy: Europe's DeCiDe Project. *ASTC Dimensions* (September-October), 10-11.

Barab, S. & Squire, K. (2004): Design-Based Research: Putting a stake in the ground, *The Journal of the Learning Sciences*, 13 (1), 1-14.

Barnett, R. (2004). Learning for an unknown future. *Higher Education Research and Development*, 23(3), 247-260.

Baumgartner, E. & The Design Based Research Collective (2003): Design-Based Research: An emerging paradigm for educational inquiry, *Educational Researcher*, 32 (1), 5-8.

Bradburne, J.M. 2001. Going public: Science museums, debate and democracy. *The Review of Education, Pedagogy, and Cultural Studies* 23 (4), 349-379.

Bruner, J. (1986): *Actual minds, possible worlds*. London: Harvard University press.

Bruner, J. (1990): *Acts of meaning*. Cambridge & London: Harvard University Press.

Bruner, J. (1991): The narrative construction of reality. *Critical Inquiry*, 18(1), 1-21.

Bruner, J. (1996): *The culture of education*, London: Harvard University Press

Buckingham, D. and M. Scanlon. (2003). *Education, Entertainment and Learning in the Home*. Philadelphia: Open University Press.

Csikszentmihalyi, M. (1990): *Flow: The psychology of optimal experience*. New York: Harper Collins

Csikszentmihalyi, M. & Hermanson, K. (1995): Intrinsic motivation within museums: Why does on want to learn? In J. Falk & L. Dierking (eds.), *Public institutions for personal learning: Establishing a research agenda*, pp. 67 -77, Washington D.C.: American Associations of Museums

Flick, U. (2006): *An introduction to qualitative research*. London: SAGE Publications.

Gergen, K. J. (2002). *Realities and relations: Soundings in social construction*: Harvard University Press.

Goffman, E. (1959). *The presentation of self in everyday life*. New York: Doubleday.

Henriksen, E. K., & Frøyland, M. (2000). The contribution of museums to scientific literacy: Views from audience and museum professionals. *Public Understanding of Science*, 9, 393-415.

Hooper-Greenhill, E. (ed.) (1999): *The educational role of the museum* (2nd Ed.). London: Rutledge.

Jordan, B., & Henderson, A. (1995). Interaction analysis: Foundations and practice. *The Journal of the Learning Sciences*, 4(1), 39-103.

Kahr-Højland, A. 2007. The mobile phone as a museum piece? Mobiles boding for a paradigm shift required in the learning museum anno 2007. In Philipsen, H. and L. Qvortup L (eds.) *Moving media studies - remediation revisited*, pp. 121-143. Odense: Samfundslitteratur Press.

Kahr-Højland, A. (2010a): EGO-TRAP: A Mobile Augmented Reality Tool for Science Learning in a Semi-formal Setting, *Curator*, 53 (4), 1-9

Kahr-Højland, A. (2010b): *Learning is not a joke: A study of young people's experiences with a mobile-facilitated narrative in a scientific context*, Ph.D. thesis, Department of Media Science, University of Southern Denmark.

Kvale, S. (1997): *InterView: En introduktion til det kvalitative forskningsinterview.* København: Hans Reitzels Forlag.

Labov, W (1972): *Sociolinguistic patterns.* Philadelphia: University of Pensylvania Press.

Mandler, J. M. (1984): *Stories, scripts and scenes: Aspects of schema theory*, Hillsdale, NJ: Lawrence Erlbaum Associates.

Nowotny, H., P. Scott, and M. Gibbons. 2003. "Mode 2" revisited: The new production of knowledge. *Minerva* 41, (3), 179-94.

Oppenheimer, F. (1968): Rationale for a science centre, *Curator: The Museum Journal*, 11 (3), 206-209 (September 1968).

Paris, S. 1997. Situated motivation and informal learning. *Journal of Museums Education* 22, (2-3), 22-27.

Parry, R. 2007. *Decoding the museum: Digital heritage and the technologies of change.* London: Routledge.

Pedretti, E. 2004. Perspectives on learning through critical issues-based science centre exhibits. *Science Education* 88 (1), 34-47

Quistgaard, N. & Kahr-Højland, A. (2010). New and innovative exhibition concepts at science centres using communication technologies. *Museum Management and Curatorship*, 25(4), 423-436.

Quistgaard, N. 2006. Oplevelsen og udbyttet af skolebesøg pa teknik- og naturvidenskabscenter. *MONA* 1, 23-40.

Qvortrup, L. (2010): *Autoriteten - den tredje vej*, Interview Eksistens, DR Broadcast, july 25th 2010 URL: http://www.dr.dk/P1/Eksistens/20050902132728.htm

Rennie, L.J., and T. McClafferty. 1996. Science centres and science learning. *Studies in Science Education* 27, 53-58.

Walker, K. (2008): Structuring visitor participation. In K. Walker & L. Tallon (eds), *Digital technologies and the museum experience* (pp. 109-124). Plymouth: Altamira Press.

Wenger, Etienne (2003): Communities of practice and social learning systems. In: D. Nicoline, S. Gherardi & D. Yanow (eds.), *Knowing in organizations: A practice-based approach* (pp. 75-99), New York: M.E. Sharpe Inc.

Wood, D., Bruner, J. & Ross, G. (1976): The role of tutoring in problem solving. *Journal of child psychology and psychiatry*, 17, 89-100.

Mobile Devices for Promoting Museum Learning

TIMOTHY D ZIMMERMAN

Rutgers University

"Museums and libraries were my teachers. I didn't like school very much." Oliver Sachs, 2008

When we think of visiting a museum, aquarium, or historical site, we often think of fun times with family, friends, or maybe time away from school. But many people, including great thinkers such as Oliver Sachs, see museums as places of learning. Indeed, schools in the U.S. and other countries include field trips to museums as part of students' schooling experience. However, the connections between what is taught in school and what is taught/experienced in museums (or other informal settings) are often tenuous at best. Indeed, researchers have called for developing a better understanding of how students connect ideas between museums and classrooms (Knutson, 2007, Martin, 2007). Existing research on connecting school and museum learning has focused on pre-visit learning activities, the link between museum visits and follow-up activities, and to some degree, the role of facilitators during the visit. However, very few studies have focused on facilitating learning over the entire sequence of in-school preparation – museum visit – in-school concluding activities. In this chapter, I focus on aspects of learning in museums and argue that mobile devices, in particular, can be leveraged to facilitate learning in and across museum and school (or home, multiple museums, playground, etc.) settings, an approach I call *Learning Across Contexts.*

Background

Why study learning in (or connected to) museums?

One answer to this question is simply one of scale: lots of people visit museums in the U.S. To put the number of visitors in perspective, consider the results of a survey by the Association of Zoos and Aquariums (AZA). With responses from one set of surveys, they estimated that in excess of 121 million people visit U.S. zoos and aquariums each year (AZA, 1999). That is about the same number of people attending all professional spectator sporting events in the U.S. combined (NBA plus NFL plus NASCAR plus MLB – you get the picture) (U.S. Census Bureau, 2000). In addition to scale, people generally perceive visits to museums as learning experiences. In fact, in poll after poll, visitors to museums state that their first object for visiting is to have fun and their second (a very close second) objective is to learn (McIntyre, 2007). Finally, studying learning in museums, as it is connected to school learning, has emerged as a new area of research in recent years, creating new opportunities for investigating technology in education (c.f., Bell et al., 2009; DeWitt and Osborne, 2007; Zimmerman, 2009, 2010).

Although the different types of institutions that provide opportunities for out-of-school learning are diverse (including zoos, aquariums, and after-school program settings), these institutions share many similar features. Thus, for consistency, I use the term *museum* to refer to such institutions and the term *informal learning* to refer to the cognitive, affective, and attitudinal changes resulting from visits thereto.

Many scholars have attempted to differentiate school-based (a.k.a. *formal*) and museum-based (often referred to as *informal*) learning. Somme have created lists and tables articulating how learning itself differs between formal and informal learning settings. More recently, Falk and Dierking (1992; 2000) proposed the Contextual Model of learning, which addresses the range of learning that occurs in what they call free-choice learning environments. Falk and Dierking claim that learning is always a highly personal process that is strongly dependent upon prior knowledge and experiences, and occurs within unique sociocultural and physical contexts. They argue that learning involves multiple sources of experience and information, which collectively contribute to knowledge construction. This focus on context as the appropriate unit of analysis led the Informal Science Education Ad Hoc Committee of the Board of the National Association of Research in Science Teaching to conclude that *learning* is essentially the same in both settings, but the *context* is what differs (Dierking, et al., 2003). Learning that occurs in a classroom context typically involves long, sequenced units that rely on previously learned scientific concepts, formalized testing, teacher-student hierarchies, and teacher (or state) dictated content. Museum-based learning, however, often occurs in short time units discontinuously across time and space, often with heterogeneous groups of students; it does not require specific sequences of content, and is based on curiosity, intrinsic motivation, and student choice of topic, with little or no evaluative component (Csikszentmihalyi and Hermanson, 1995; Dierking, et al., 2003; Falk and Dierking, 1992, 2000; Griffin and Symington, 1997; Bell, et al., 2009; Pedretti, 2002; Rennie, 2007; Rennie and McClafferty, 1996). Learning in museums results from unique social interactions, hands-on experiences, and free-choice opportunities that are afforded by the museum context (Allen, 1997; Ash, 2002; Bell, et al., 2009; Ellenbogen, 2002; Falk and Storksdieck, 2005; Hein, 1998).

Informal learning experiences are often a part of school-based learning (such as field trips to museums, zoos, aquariums, or outdoor learning spaces), have long been a part of American culture, and are even considered "a good, enriching, educational activity" that contributes to the achievement of scientific literacy (NRC, 1996). Research on school-based field trips to museums has increased in recent years and good reviews have been published by Griffin (2004) and DeWitt and Storksdieck (2008). In short, much of this research focuses on the educational value of the class visit (Bamberger and Tal, 2007; Cox-Petersen et al., 2003; Griffin, 2004, Pedretti, 2004), the mediating role of adults attending such field trips (Sedzielarz, 2003; Tal and Steiner, 2006; Tal, Bamberger and Morag, 2005), effective preparation activities/strategies (Anderson, Lucas, and Ginns, 2003; Lebeau, et al., 2001; Lucas, 2000; Orion, 1993; Tal and Morag, 2007), or post-trip wrap-up activities (Anderson, et al., 2000). Importantly, there is still a gap in the literature around understanding the nature of learning across these contexts, leading some to call for research on understanding how students connect ideas across these contexts, and pedagogical approaches that effectively foster learning connections across contexts (Knutson, 2007; Martin, 2007; Bell, et al., 2009; Rennie, 2007). A few such studies now exist (c.f., Kali, Orion and Eylon, 2003; Zimmerman, 2010) but currently only Zimmerman (2005, 2010) has leveraged mobile technology as a tool for connecting learning across formal and informal contexts, which has yielded some evidence of this continuity of learning (Zimmerman, 2009, 2010). As will be described below, mobile technology provides important scaffolding to facilitate across context learning.

Studying and promoting learning across contexts through mobile technologies

The Ocean Sciences Education Research group at Rutgers University (OSER – http://oser.gse.rutgers.edu) is working to bring together educational research within and across school and museum contexts, desktop and mobile computer technologies, internet connectivity, and the learning and teaching of ocean science content. This has resulted in a three-phase pedagogical approach to museum-based field trips, an approach that is promoted by other researchers (e.g., DeWitt and Osborne, 2007; Storksdiek, 2006) and made possible by mobile technologies. We call this pedagogical approach *Learning Across Contexts*. To achieve this, we meld together various theoretical perspectives, multiple educational research and museum studies methodologies, and cutting-edge technology. We embed all these in a constructivist approach to learning, a commonly stated view of learning by learning context designers in classrooms and museums (c.f., Bransford, et al., 2000; Hein, 1998).

Perspectives on learning and teaching

In general, constructivism views learning as an active cognitive process of knowledge building, reliant upon prior ideas, experiences, and understandings, that is socially, culturally, and contextually influenced (Bransford, et al., 2000; Piaget and Inhelder, 1963; Vygotsky, 1978). This notion of learning is promoted not only by those associated with the formal (school-based) learning community, but by the learning museum community as well (Falk and Dierking, 1992; Hein, 1998). A central component of most constructivist approaches is that the learner is the focus of teaching and learning, and not the content or the teacher (Bransford, et al., 2000; Lemke, 1990). The museum setting provides a context where the locus of control lies squarely with the learner, and not the (perceived) more knowledgeable "teacher" (docent, explainer, chaperone, parent), allowing for a natural learner-centered experience (Hein, 1998).

In recent years, sociocultural learning theory became widely used by researchers for describing and explaining learning, especially during the study of museum visits. This theoretical lens allows researchers and museum professionals to identify a variety of interactions among children and adults related to their prior experiences in other settings (Allen,

1997; Ash, 2003; Leinhardt, et al., 2002). This lens reveals how the museum experience is more meaningful if it is connected to the visitors' world and experiences (Lebeau, et al., 2001; Pedretti, 2002). The sociocultural context of learning takes into account interactions among group members (families, school groups) and between group and out-of-group members (museum personnel). Museums are considered strongly socioculturally mediated settings (Ash, 2004; Ash and Wells, 2006; Falk and Dierking, 1992; Rennie, et al., 2003). Studies on the mediating role of adults in museums – parents, teachers, chaperones and museum guides demonstrate the importance of adult-young learner interactions in these contexts (Ash, 2004; Tal, Bamberger and Morag, 2005; Cox-Petersen et al., 2003; Crowley et al., 2001; Sedzielarz, 2003) – while other studies investigated students' conversations and collaborative learning in museums during class visits points to the important pedagogical role these people play during visits (Kisiel, 2003; Lebeau et al., 2001; Priest and Gilbert, 1994). The OSER group has recently applied both constructivist and socio-cultural lenses to research on mobile technology use across learning contexts by designing new mobile learning tools and by recording students' conversations during school-based field trips to an aquarium. While still in the early stages, this research is already yielding interesting results on the use of mobile technology to facilitate learning in museums (Zimmerman, 2010).

From a pedagogical perspective, experiential (hands-on), learner-centered, and inquiry-based curricula have emerged as productive approaches to teaching science (NRC, 1996, 2000; 2005; Davis, 2000; White and Frederickson, 1998; Zimmerman and Stage, 2008). Though there is some dispute over the effectiveness of inquiry-based pedagogical approaches (c.f., Kirshner, et al., 2006; Hmelo-Silver, et al., 2007), Hmelo-Silver, et al. (2007) draw upon considerable empirical evidence to argue for the use of inquiry in classrooms. Though this evidence is focused on inquiry in the classroom, this definition of inquiry, one that views science learning as a complex treatment of both science content and science inquiry practices (Hmelo-Silver, et al., 2007), is recognized as important for learning in museums as well (c.f., Allen, 2004; Bell, et al., 2009; Gutwill and Allen, 2010; Randol, 2005; Zimmerman, 2010).

Knowledge integration view of learning and teaching

Knowledge integration (Linn, Eylon, and Davis, 2004; Linn and Hsi, 2000) refers to a specific view of inquiry-based, constructivist learning. Knowledge integration is a process of learning that incorporates the addition of concepts, ideas, experience, beliefs, and other knowledge constructs into one's *repertoire of ideas* with the development and sorting out of connections among and between knowledge constructs to form a *knowledge network* (Linn, et al., 2004). This perspective on learning includes both the cognitive processes and natural learning activities in which we engage throughout our lives. Learners make observations, interpret them, and are deliberate in their own learning, all within social and cultural contexts. They also add, promote, and demote ideas, and make or rearrange connections to ideas while evaluating their merit (Linn, et al., 2004; Linn and Hsi, 2000). Since inquiry-oriented curricula provide a context for revisiting prior ideas, constructing new ideas, and linking ideas, a knowledge integration perspective on learning complements the use of inquiry curricula and is used in OSER research.

Learning across contexts pedagogical approach

The Learning Across Contexts (LAC) curriculum design approach draws from classroom-based and informal contexts, research and theoretical perspectives, and advocates integrated learning experiences across contexts by applying the knowledge integration perspective to the curriculum design and research approach. LAC seeks to promote deeper science learning and provide a research context to better understand the nature of learning across various formal and informal learning contexts through a combination of:

1. new/emerging technologies;
2. inquiry-based approaches to science learning; and
3. school-led field trips, self- or docent-directed outdoor and environmental education experiences, or citizen science research projects.

This work draws heavily from the learning sciences literature, especially as it relates to design experiment approaches to educational research (Confrey, 2006), the use of technology for teaching and learning (Sawyer, 2006), and multiple

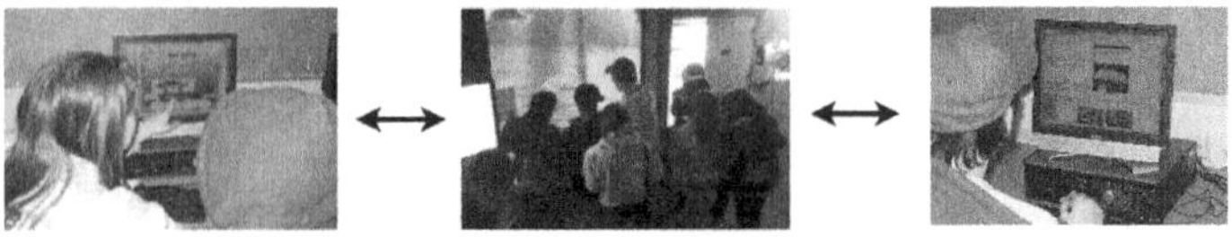

Figure 1: The three-phase LAC Curriculum Design approach. Phase 1 is the pre-field trip phase where students work together on classroom computers to prepare for the field trip. Phase 2 is the visit to the informal learning context where students interact with each other in the real world and virtual world (using mobile devices). Phase 3 is the post-field trip phase where students again work together on classroom computers reflecting on learning during phases 1, 2 and 3.

research methods for understanding learning in naturalistic settings.

In order to provide context for LAC and address concerns identified in the research literature around the need for preparatory and follow-up activities (Anderson, et al., 2000; DeWitt and Osborne, 2007; Orion and Hofstein, 1994), the informal learning experience is designed to be integral to the curriculum, not just an added activity. To achieve this, a three-phase pedagogical design (Zimmerman, 2005, 2009; Zimmerman, 2010) was developed (Fig. 1 above) with the use of mobile devices playing the central role of scaffolding the museum visit.

Integrating technologies to promote learning across contexts

In recent years, educational researchers have successfully used a combination of educational technologies for engaging students in inquiry-based science learning activities (Slotta and Linn, 2009). For example, the Web-based Inquiry Science

Environment (WISE – see http://wise.berkeley.edu) offers numerous inquiry tools such as drawing, graphing, data tables, concept mapping, online discussions, and student journals to support student learning in classroom settings (Linn and Hsi, 2000; Linn and Slotta, 2000; Linn, Clark, and Slotta, 2003; Slotta, 2004; Slotta and Linn, 2000, 2009). WISE provides a powerful, convenient platform for the design and delivery of interactive inquiry projects that engage students with internet resources and encourage them to reflect and make connections to real world experiences and ideas.

The use of handheld (mobile) devices in educational settings has been around as long as the devices themselves. For example, Soloway and colleagues (Soloway, Norris, and Blumenfeld, 2001) applied the principles of learner-centered design to develop handheld research scaffolds and data collection tools for Highly Interactive Computing in Education (http://umich.edu/~hiceweb/). These designs included game-like activities and web-like content often designed for use on handheld devices that required synching to transfer data and software between the device and a desktop or laptop computer. This form factor was common in the early educational use of mobile devices. In another example, Nancy Songer and colleagues developed BioKIDS, a powerful application of handheld computers that support elementary students' playground investigations of animal biodiversity (Parr, et al., 2004; Songer, 2006). With BioKIDS, designers employed iconic depictions and a dichotomous key, similar to what science researchers use for identifying species, to develop a tool that allowed elementary-school students to correctly identify and count animal species on the school grounds. This tool was delivered to kids via PDAs. In this way, the BioKIDS project sought to emulate authentic science practice through the use of mobile devices. At the same time, the WISE research group at UC Berkeley developed opportunities for handheld use both inside and outside classrooms. This work involved data collection activities and game-like simulations on Palm IIIc handhelds that were interconnected with activities in the WISE learning environment. Together, educational technologists, teachers, educational researchers, and museum staff developed HTML data collection forms that were delivered to students on the Palm IIIc mobile devices. This allowed the mobile devices to work as both data collection tool and learning facilitator. For example, during a school field trip to an aquarium, the handhelds provided a tool for data collection, cognitive and metacognitive scaffolds such as field guides, hints to reduce cognitive load, and metacognitive reflection notes to prompt connections of ideas between the classroom and aquarium (Aleahmad and Slotta, 2002; Slotta and Zimmerman, 2002, 2004; Zimmerman and Slotta, 2003) (See Fig. 2 below).

As the use of mobile devices became more widely used in formal settings, their importance in non-school-based learning experiences began to rise. For example, Hsi (2003) studied visitors' experiences using "nomadic" devices to facilitate their experience at a hands-on interactive science museum. Designers sought to both increase visitors' experience and

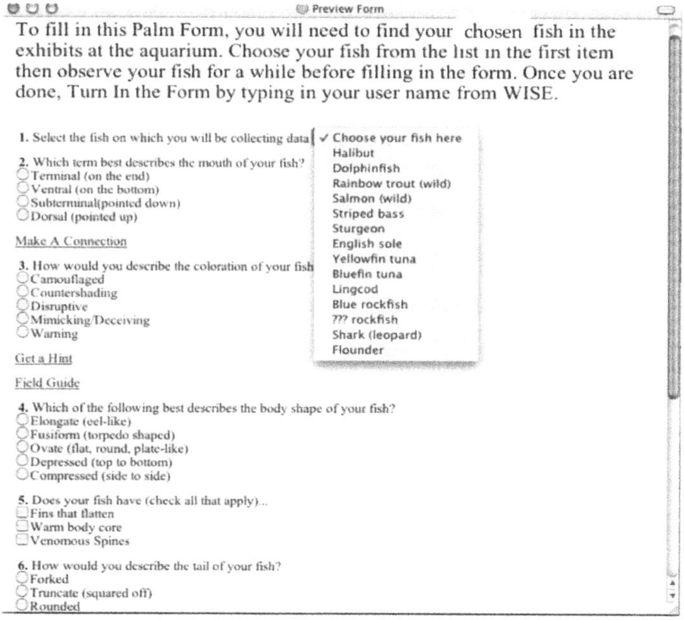

Preview Form

To fill in this Palm Form, you will need to find your chosen fish in the exhibits at the aquarium. Choose your fish from the list in the first item then observe your fish for a while before filling in the form. Once you are done, Turn In the Form by typing in your user name from WISE.

1. Select the fish on which you will be collecting data

✓ Choose your fish here
Halibut
Dolphinfish
Rainbow trout (wild)
Salmon (wild)
Striped bass
Sturgeon
English sole
Yellowfin tuna
Bluefin tuna
Lingcod
Blue rockfish
??? rockfish
Shark (leopard)
Flounder

2. Which term best describes the mouth of your fish?
- Terminal (on the end)
- Ventral (on the bottom)
- Subterminal(pointed down)
- Dorsal (pointed up)

Make A Connection

3. How would you describe the coloration of your fish
- Camouflaged
- Countershading
- Disruptive
- Mimicking/Deceiving
- Warning

Get a Hint

Field Guide

4. Which of the following best describes the body shape of your fish?
- Elongate (eel-like)
- Fusiform (torpedo shaped)
- Ovate (flat, round, plate-like)
- Depressed (top to bottom)
- Compressed (side to side)

5. Does your fish have (check all that apply)...
- Fins that flatten
- Warm body core
- Venomous Spines

6. How would you describe the tail of your fish?
- Forked
- Truncate (squared off)
- Rounded

Figure 2: Screenshot of an HTML mobile form created for student use during an aquarium field trip. The items underlined are hyperlinks that open new HTML pages containing either a Field Guide (containing picture, science terms and explanatory text), Make A Connection prompts (questions designed to promote student reflection back to classroom experiences or forward to future class work), or Get a hint (Socratic-style questions that help students conduct Data Collection activities).

provide enhanced learning opportunities through station-based RFID delivery of web content and learning scaffolds. As the number, size, and ubiquity of mobile devices increased, so has their use in museum settings. Phipps, Rowe, and Cane (2009) harnessed the ever-increasing capabilities of the iPod. Using prerecorded audio and video content, they provided visitors with additional ways to experience exhibits at a marine science visitors' center. In the past several years, as wireless and cellular networks have become more accessible, the pace of development has increased in the area of handheld use for education both within classrooms and for learning outside of classrooms. For example, the RAFT project (Kravcik et al., 2004) allows students and teachers to collaborate in real-time using internet connectivity while on a field trip. Another example is the use of mobile devices as part of citizen science projects, where "average citizens" help scientists collect valuable data, such as bird monitoring (see Chen et al., 2002). Many other researchers have used mobile devices in education inside classroom and museums (e.g., Klopfer, et al., 2004; Kurti et al., 2006; Roschell and Pea, 2002), but often the activities are not tied to a larger curriculum and thus do not attempt to harness the devices' capabilities as facilitators of across-context learning.

The LAC pedagogical approach seeks to help learners connect learning that occurs in and across museum and school (or home, bus, playground) contexts through internet-connected mobile devices. We are currently creating OSER LAC curricula that scaffold learners' visits to the museum with carefully designed observation guides, prompts, and tools – which allow students to store, sort, tag, photograph, post, and analyze information as they move from context to context. These will allow students to work collaboratively on mobile or desktop computers, prepare for a museum (or other informal learning environment) experience, reflect on science concepts from multiple contexts, and become sophisticated users of science across various temporal and spatial scales.

As we move forward with mobile technology designs for museums, we need to think critically about helping learners, whether currently in school or long out of school, connect what they already learned elsewhere to what they are learning in the museum. The LAC pedagogical model seeks to accomplish this with school-led field trips, but the implications go beyond the school associated activities.

Moving forward

Over 100 years ago, John Dewey envisioned a school-museum institution (Dewey, 1900). A physical structure that would contain aspects of a typical school embedded within the structural components of a museum. This vision emerged from Dewey's philosophy of education, particularly his notion of continuity, the idea that "every experience both takes up something from those which have gone before and modifies in some way the quality of those which come after," which gave rise to the idea that experiences in museums are integral to learning that happens in school, or elsewhere (Dewey, 1916, p. 35). However, no longer do we need to think in terms of building physical structures or even singular institutions to foster a continuity of learning; we now have technology that can scaffold that connection for learners. Imagine, a family (or a school group, a couple, a group of friends) decides to attend a science museum one weekend. While checking the museum's web site, they encounter a pop-up box asking if they want to use the museum's mobile app to prepare for, record, and share their museum visit. After a quick login and mobile download, the members of the group plan their visit, answer short questions about what they expect, and even post their trip on a social networking site. While visiting, they view their answers, make changes, and make connections to previous (or other collaborating museum) experiences. Upon returning home, a text message prompts the visitors for final reflections, posting of photos, wiki creation about an upcoming exhibit, and contribution to the online database. This is all now possible, and will only improve as technology improves. With this scenario in mind, I'd like to end with what I hope will be some provocative questions.

- How do we extend learning by leveraging museums, mobile devices and social networking?
- What kinds of mobile tools are needed to promote learning in museums, but also across museum and non-museum learning spaces?
- How do we leverage 24/7/365 connectivity to reinvent Dewey's vision of seamless museum-school learning experiences?
- Can technology help us extend "learning" beyond the school walls, beyond the museum walls, to foster lifelong and life-wide learning for all who wish to learn?

References

Alehamad, T., & Slotta J.D. (2002, June). *Integrating Handheld Technology and Web-based Science Activities.* Paper presented at the meeting of ED-MEDIA, Denver, CO.

Allen, S. (2004). Designs for learning: Studying science museum exhibits that do more than entertain. *Science Education*, 88(Suppl. 1), S17–S33.

Allen, S. (1997). Using scientific inquiry activities in exhibit explanations. *Science Education*, 81(6), 715-734.

Anderson, D., Lucas, K. B., Ginns, I.S. (2003). Theoretical perspectives on learning in an informal setting. *Journal of Research in Science Teaching*, 40(2), 177-199.

Anderson, D., Lucas, K.B., Ginns, I.S., & Dierking, L.D. (2000). Development of knowledge about electricity and magnetism during a visit to a science museum and related post-visit activities. *Science Education*, 71, 658-679.

Ash, D. (2002). Negotiation of biological thematic conversations in informal learning settings. In G. Leinhardt, K. Crowley, and K. Knutson (Eds.), *Learning conversations in museums* (pp. 357 – 400). Mahwah, NJ: Lawrence Erlbaum

Ash, D. (2003). Dialogic inquiry in life science conversations of family groups in a museum. *Journal of Research in Science Teaching*, 40, 138–162.

Ash, D. (2004). *Reflective scientific sense-making dialogue in two languages: The science in the dialogue and the dialogue in the science*, 88(6), 855-884.

Ash, D. & Wells, G. (2006). Dialogic inquiry in classroom and museum: Actions, tools, and talk. In Z. Bekerman, N.

Burbules, & D. Silberman-Keller (Eds), *Learning in places: The informal education reader* (pp. 35-54). New York, NY: Peter Lange.

Association of Zoos and Aquariums (AZA). (January 1999). *The Collective Impact of America's Zoos and Aquariums*. AZA: Silver Spring, MD.

Bamberger, Y. & Tal, T. (2007). Learning in a personal context: Levels of choice in a free choice learning environment in science and natural history museums. *Science Education*, 91(1), 75-95.

Bell, P., Lewenstein, B., Shouse, A.W., & Feder, M.A. (Eds.). (2009). *Learning science in informal environments: People, places, and pursuits*. Washington, D.C.: National Academies Press.

Bransford, J.D., Brown, A.L., & Cocking, R.R. (2000). *How people learn: Mind, experience, and school*. Washington, DC: National Academy Press.

Chen, Y. S., Kao, T. C., Sheu, J. P., & Chiang, C. Y. (2002). *A mobile scaffolding aid-based bird watching learning system*. Paper presented at Wireless and Mobile Technologies in Education, Växjö, Sweden.

Confrey, J. (2006). The evolution of design studies as methodology. In Sawyer, RK (Ed.), *The Cambridge handbook of the learning sciences*. (p. 135-151). Cambridge, UK: Cambridge University Press.

Cox-Petersen, A.M., Marsh, D.D., Kisiel, J., & Melber, L.M. (2003). Investigation of guided school tours, student learning, and science reform recommendations at a museum of natural history. *Journal of Research in Science Teaching*, 40(2), 200-218.

Csikszentmihalyi, M. & Hermanson, K. (1995). Intrinsic motivation in museums: What makes visitors want to learn? In: J.H. Falk & L.D. Dierking (Eds.) *Public institutions for personal learning: Establishing a research agenda*, pp. 67-77. Washington, D.C.: American Association of Museums.

Davis, E. A. (2000). Scaffolding students' knowledge integration: Prompts for reflection in KIE. International *Journal of Science Education*, 22(8), 819-837.

Dewey, J. (1900). *The School and Society*. Chicago: University of Chicago Press.

Dewey, J. (1916). *Democracy and education: An introduction to the philosophy of education*. New York: NY, The Macmillan Company.

DeWitt, J., & Storksdieck, M. (2008). A short review of school field trips: Key findings from the past and implications for the future. *Visitor Studies*, 11,181–197.

DeWitt, J., & Osborne, J. (2007). Supporting teachers on science-focused school trips: towards an integrated framework of theory and practice. *International Journal of Science Education*, 19, 685–710.

Dierking, L.D., Falk, J.H., Rennie, L., Anderson, D., & Ellenbogen, K. (2003). Policy statement of the "Informal Science Education" Ad Hoc Committee, *Journal of Research in Science Teaching*, 40, 108-111.

Ellenbogen, K. (2002). Museums in family life: An ethnographic case study. In Leinhardt, G., Crowley, K.,

& Knutson, K. (Eds). *Learning conversations in museums.* Mahwah,NJ: Erlbaum. (p. 81-102).

Falk, J. H., & Dierking, L. D. (2000). *Learning from museums: Visitor experiences and the making of meaning.* Walnut Creek, CA: AltaMira Press.

Falk, J.H., & Dierking, L.D. (1992). *The museum experience.* Washington, DC: Whalesback.

Falk, J. H. & Storksdieck, M. (2005). Using the Contextual Model of Learning to understand visitor learning from a science center exhibition. *Science Education*, 89(5), 744 - 778.

Griffin, J. (2004). Research on students and museums: Looking more closely at the students in school groups. *Science Education*, 88(suppl), S59-S70.

Griffin, J., & Symington, D. (1997). Moving from task-oriented to learning-oriented strategies on school excursions to museums. *Science Education*, 81(6), 763-779.

Gutwill, J.P., & Allen, S. (2010). Facilitating family group inquiry at science museum exhibits. *Science Education*, 94(4), 710-742.

Hein, G. (1998). *Learning in the museum.* New York, NY: Routledge.

Hmelo-Silver, C.E., Duncan, R.G & Chinn, C.A. (2007). Scaffolding and Achievement in Problem-Based and Inquiry Learning: A Response to Kirschner, Sweller, and Clark (2006). *Educational Psychologist*, 42(2): 99-107.

Hsi, S. (2003). A study of user experiences mediated by nomadic web content in a museum. *Journal of Computer*

Assisted Learning, 19(3), 308–319.

Kali, Y., Orion, N., & Eylon, B. (2003). Effect of knowledge integration activities on students' perception of the Earth's crust as a cyclic system. *Journal of Research in Science Teaching*, 40(6), 545-565.

Kirschner, P.A., Sweller, J., & Clark, R.E. (2006). *Why Minimal Guidance During Instruction Does Not Work: An Analysis of the Failure of Constructivist, Discovery, Problem-Based, Experiential, and Inquiry-Based Teaching*, 41(2), 75-86.

Kisiel, J. (2003). Teachers, museums and worksheets: A closer look at learning experiences. *Journal of Science Teacher Education*, 14(1), 3-21.

Klopfer, E., Yoon, S., & Perry, J. (2004). Using Palm technology in participatory simulations of complex systems: A new take on ubiquitous and accessible mobile computing. *Journal of Science Education and Technology*, 14(3), 285-297.

Knutson, K. (2007, June). *Interview with Professor Richard Duschl about an agenda for informal science education and research in scientific reasoning and argumentation.* Interviews in the Field. Retrieved June 30, 2007, from http://www.informalscience.org/community/interview/richardduschl.html.

Kravcik M., Kaibel, A., Specht, M., and Terrenghi, L. (2004). Mobile Collector for Field Trips. *Educational Technology & Society*, 7 (2), 25-33.

Kurti, A., Milrad, M., Alserin, F., & Gustafsson, J. (2006) Designing and implementing ubiquitous learning activities supported by mobile and positioning technologies. In V. Uskov (Ed), *Proceedings of the Ninth IASTED International Conference on Computers and Advanced Technology in Education*. Anaheim, CA: IASTED.

Lebeau, R.B., Gyamfi, P., Wizevich, K., & Koster, E.H. (2001). Supporting and documenting choice in free-choice science learning environments. In J.H. Falk (Ed.), *Free-choice science education: How we learn science outside of school*. New York, NY: Teachers College Press.

Leinhardt, G., Crowley, K., & Knutson, K. (2002). *Learning conversations in museums*. Mahwah,NJ: Erlbaum.

Lemke, J. L. (1990). *Talking science: Language, learning, and values*. Norwood, NJ: Ablex Publishing Corporation.

Linn, M. C., & Hsi, S. (2000). *Computer, teachers, peers: Science learning partners*. Mahwah, NJ: Erlbaum

Linn, M.C. & Slotta, J.D. (2000). WISE Science. *Educational Leadership*, 52(2), 29-32.

Linn, M. C., Clark, D., & Slotta, J. D. (2003). *WISE design for knowledge integration*, 87(4), 517-538.

Linn, M.C., Eylon, B., & Davis, E.A. (2004). *The knowledge integration perspective on learning*. In M.C. Linn, E.A. Davis, & P. Bell (Eds.).

Lucas, K. B. (2000). One teacher's agenda for a class visit to an interactive science teacher. *Science Education*, 84, 524–544.

Martin, LMW. (2007). An Emerging Research Framework for Studying Free-Choice Learning and Schools. In J.H. Falk, L.D. Dierking, and S. Foutz (Eds.), *In Principle, In Practice: Museum as Learning Institutions* (pp.247-259). Plymouth,

UK: AltaMira Press.
McIntyre, M.H. (2007). *Audience knowledge digest: Why people visit museums and galleries, and what can be done to attract them*. Retrieved from: research.mla.gov.uk/evidence/download-publication.php?id=951 on Oct. 25, 2010.
National Research Council (NRC). (1996). *National Science Education Standards*. Washington, DC: National Academy Press.
National Research Council (NRC). (2000). *Inquiry and the national science education standards: A guide for teaching and learning*. Washington, D.C.: National Academy Press.
National Research Council (NRC). (2005). *How students learn history, mathematics, and science in the classroom*. Washington, D.C.: The National Academy Press.
Orion, N. (1993). A model for the development and the implementation of field trips as an integral part of the science curriculum. *School Science and Mathematics*, 93(6), 325-331.
Orion, N., & Hofstein, A. (1994). Factors that influence learning during a scientific field trip in a natural environment. *Journal of Research in Science Teaching*, 31, 1097-1119.
Parr, C.S., Jones, T.J. & Songer, N.B. (2004). Evaluation of a Handheld Data Collection Interface for Science Learning. *Journal of Science Education and Technology*, 13(2), 233-242.
Pedretti, E.G. (2004). Perspectives on learning through research on critical issues-based science center exhibitions. *Science Education*, 88 (suppl), S34-S47.
Pedretti, E. (2002) T. Kuhn meets T. Rex: critical conversations

and new directions in science centers and science museums. *Studies in Science Education*, 37, 1–42.

Piaget, J., and Inhelder, B. (1962). *The Psychology of the Child*. New York: Basic Books

Priest, M., & Gilbert, J. (1994). Learning in museums: Situated cognition in practice. *Journal of Education in Museums*,15,16–18.

Randol, S.M. (2005) *The nature of inquiry in science centers: Describing and assessing inquiry at exhibits*. Unpublished Doctoral Dissertation, University of California, Berkeley.

Rennie, L.J. (2007). Learning science outside of school. In: S.K. Abell & N.G. Lederman (Eds.), *Handbook of research on science education* (pp. 125-167). Mahwah, NJ: Lawrence Erlbaum Assoc., Inc.

Rennie, L., & Mcclafferty, T. (1995). Using visits to interactive science and technology centers, museums, aquaria, and zoos to promote learning in science. *Journal of Science Teacher Education*, 6(4), 175-185.

Rennie, L., Feher, E., Dierking, L., & Falk, J. (2003). Toward an agenda for advancing research on science learning in out-of-school settings. *Journal of Research in Science Teaching*, 40, 112–120.

Roschelle, J., & Pea, R. (2002). A walk on the WILD side: How wireless handhelds may change computer-supported collaborative learning. *International Journal of Cognition and Technology*, 1(1), 145–168.

Sachs, O. (2008). *Creativity and the brain*. Talk presented at the Writers at Rutgers Reading Series, Rutgers University, New

Brunswick, NJ.

Sawyer,R.K. (2006). Introduction: The new science of learning. In R.K. Sawyer (Ed.), *The Cambridge Handbook of the Learning Sciences*, (pp. 1-18). New York: Cambridge University Press.

Sedzielarz, M. (2003). Watching the chaperones: An ethnographic study of adult-child interactions in school field trips. *Journal of Museum Education* 28(2): 20-24.

Slotta, J. D. (2004). The Web-based Inquiry Science Environment (WISE): Scaffolding Knowledge Integration in the Science Classroom. In M. C. Linn, E. A. Davis, & P. Bell (Eds.), *Internet Environments for Science Education* (pp. 203–232). Mahwah, NJ: Lawrence Erlbaum Associates.

Slotta, J.D., & Linn, M.C. (2009). *WISE science: Web-based inquiry in the classroom*. New York, NY: Teachers College Press.

Slotta, J.D., & Linn, M.C. (2000). The knowledge integration environment: Helping students use the Internet effectively. In M. J. Jacobson & R. Kozma (Eds), *Innovations in science and mathematics education: Advanced designs for technologies of learning*. Mahwah, NJ: Erlbaum, 193-226.

Slotta, J. D., & Zimmerman, T. D. (2004, April). *Integrating Handheld Activities into Web-based Inquiry Projects.* Paper presented at the annual meeting of the American Educational Research Association, San Diego, CA.

Slotta, J.D., & Zimmerman, T.D. (2002). *Designing marine science technology-mediated curricula that integrate formal and informal learning environments.* Presented at CILT Ubiquitous Computing: Handhelds in Education conference. Hillsboro, OR.

Songer, N. B. (2006). BioKIDS: An animated conversation on the development of curricular activity structures for inquiry science. In R. K. Sawyer, (Ed.) *The Cambridge Handbook of the Learning Sciences* (pp. 355-369). New York: Cambridge University Press.

Songer, N.B. (1996). Exploring learning opportunities in coordinated network-enhanced classrooms: A case of kids as global scientists. *Journal of the Learning Sciences*, 5(4), 297-327.

Storksdiek, M. (2006). *Field trips in environmental education.* Berlin, Germany:Berliner Wissenschafts-Verlag.

Tal, T. & Morag, O. (2007) School visits to natural history museums: Teaching or enriching? *Journal of Research in Science Teaching*, 44(5), 747-769.

Tal, T. & Steiner, L. (2006). Patterns of teaching-museum staff relationship: School visits to the education center of a science museum. *Canadian Journal of Science, Mathematics and Technology Education*, 6, 25-46.

Tal, T., Bamberger, Y. & Morag, O. (2005). Guided school visits to natural history museums in Israel: Teachers' roles. *Science Education*, 89, 920-935.

U.S. Census Bureau. (2000). *Section 7 Parks, Recreation, and Travel.* Retrieved from http://www.census.gov/prod/www/abs/statab1995_2000.html on 10/19/10.

Vygotsky, L. S. (1978). Mind in society: The development of higher psychological processes. Cambridge, MA: Harvard University Press.

White, B.Y., & Frederiksen, J.R. (1998). Inquiry, modeling, and metacognition: Making science accessible to all students. *Cognition & Instruction*. 16(1), 1998, pp. 3-118.

Zimmerman, T.D. (June, 2010). *Capturing Learning Across Formal and Informal Contexts*. Paper presented at the Conference of the National Association for Research in Science Teaching, Philadelphia, PA.

Zimmerman, T.D. (June, 2009). *Using mobile and classroom technologies to foster and capture learning across formal and informal contexts: A Pilot study*. Poster presented at the 8th International Conference on Computer Supported Collaborative Learning, Rhodes, Greece.

Zimmerman, T.D. (2005). *Promoting knowledge integration of scientific principles and environmental stewardship: Assessing an issue-based approach to teaching evolution and marine conservation*. Unpublished Doctoral Dissertation, University of California, Berkeley.

Zimmerman, T. D. & Slotta, J. D. (2003, April). *Helping students learn difficult biology concepts through knowledge integration activities in the classroom and at an aquarium*. Paper presented at the annual meeting of the American Educational Research Association, Chicago, IL.

Zimmerman, T.D. & Stage, E.K. (2008). Teaching science for understanding. In L. Darling-Hammond (Ed.) *Powerful learning: What we know about teaching for understanding*. (pp. 151-191) San Francisco, CA: Jossey-Bass.

Placing Our Bets: Building a National Museum's Media Strategy From Scratch

COREY TIMPSON

Canadian Museum for Human Rights

Most mobile and social media discourse tends to provide insights gleaned from both successful and failed projects. Methodologies elaborated, bright ideas concretized, projects built on new or borrowed ideas and evolved, tried, and tested. Otherwise, the discourse focuses on imagination and grand designs – commentary, opinions, and extrapolations of what could be. This chapter has a slightly different focus, for both the tested and imagined have something in common. They assume an existence of some kind, a base to build from, a foundation. What does one do, however, when the Museum for which they work has yet to be built? When there is no collection? No technical infrastructure, no media policies? When there is barely even any staff? How does an institution consider its mobile and social media strategy and even begin to build global visitor engagement when starting from an empty, blank slate? Where to begin? And how?

The Canadian Museum for Human Rights (CMHR) is Canada's first national museum to be established since 1967 and is the first to be located outside of Canada's National Capital Region of Ottawa. Currently being constructed in Winnipeg, Manitoba, the Museum will reside at the historic Forks where the Red and Assiniboine rivers meet, and where people have gathered for thousands of years.[1] With construction scheduled for completion in 2012, the CMHR is busy designing its programs, amassing its content, and developing its practices and policies. The Canadian Museum for Human Rights is mandated to explore the subject of human rights, with special but not exclusive reference to

Canada, in order to enhance the public's understanding of human rights, to promote respect for others, and to encourage reflection and dialogue.[2] It will have over 47,000 square feet of exhibition space, a 20-storey glass tower, house a global reference center, and have extensive museum programs.

Interactivity

The Canadian Museum for Human Rights' subject matter, human rights, is not only largely intangible, its accepted definition and application are continuously evolving. Considering such, the museum has often been described as an "idea museum" in that its focus is not based on a museum collection.[3] Its focus is based on the concept of human rights. Internally and among colleagues, the term "dialogue museum" has also been adopted. Referencing the last portion of the Museum's mandate, and thinking towards some relatively obscure museum design discourse (see Neil Postman, "Museum as Dialogue," *Museum News*, September/October 1990, p.58), the concept of *dialogue* has very interesting implications when considering the interaction design approach of the CMHR towards exhibition, web, and overall programming goals.

When designing museum programs to be "interactive", a specific understanding of the term "interactive" has been preferred in an effort to convey far greater meaning and consequence than might be understood from acts as simple as pressing buttons on a touch screen. Considering the notion that the most interactive experience one might have in life is a great conversation,[4] designs for developing interactive experiences around a subject matter as impassioned as human rights lean perfectly toward establishing a reciprocal relationship between the museum and the visitor; a relationship where the museum informs the visitor, but where the visitor also informs the museum – hence the dialogue created. If the experience visitors might have in a gallery was expressed as a mathematical formula, then the visitors themselves would be one variable in the formula that defines their experience. Thus, in determining a definition of "interactive" as it pertains to the CMHR's experience design, the reciprocal relationship or action is mandatory. Anything else, such as pushing buttons to access video, audio, or to reveal text, is an expression of passivity.

With sessions in 19 cities across the country, speaking with and listening to thousands of Canadians, one of the Museum's first actions was to begin the dialogue. What are human rights? What do they mean? What are people's expectations from this new national institution? These topics and many more were addressed in 2009 and early 2010 by the Canadian Museum for Human Rights Content Advisory Committee and the many Canadians who came to share and interact in these public engagement sessions. When considering the Museum's programming and presence before and after opening, as well as social media, mobile media, and new technology, the opportunity for continued dialogue is obvious. Social media and mobile technology are great facilitators of this dialogue. The CMHR can relieve physical and geographical constraints by exploiting what social and mobile media are offering – an easy way to engage conversation and foster dialogue no matter where people are, and at their convenience. Through this interactivity, the Museum will inform more visitors, and more visitors will inform the Museum.

The web

A generally accepted approach to considering the Internet in the design of the museum experience comprises the pre-during-post scenario. This scenario can be described in three steps: First, the visitor interacts with the museum digitally (via the web) before arriving – this is *pre*. Secondly, the visitor visits the museum – this is *during*. Finally, the visitor who has left the museum returns online, potentially following up on what they saw in the museum – this is *post*. The pre-during-post approach has proven very popular in museum practice, yet analysis reveals success to be somewhat unclear. The *post* aspect of the scenario has, in particular, been very difficult to prove successful unless facilitated, such as with educational programming.[5]

Considering the goals of the Museum to foster and encourage dialogue and, in effect, to be interactive, in addressing the use of the Internet or the web at the Canadian Museum for Human Rights an alternate approach to the pre-during-post scenario has been preferred.

In an effort to understand the strategy that would be adopted, the concept of Web 2.0 was examined. Web 2.0 is a term that was coined by Tim O'Reilly when O'Reilly Media and MediaLive held a conference in 2004 simply titled *Web 2.0*.[6]

This term has reached mass adoption and is generally used by most individuals and institutions to express the accommodation of user collaboration, personalization, and information exchange via the web.

Software versioning is the practice by which a number or name is given to a product release. New numbers or names are used to designate new developments in software or hardware.[7] As such, if there is a Web 2.0, by definition there must have been a Web 1.0. What is the difference between the two versions? Is a Web 3.0 lurking in the not-too-distant future that the CMHR should plan for? To determine the answers to these questions, and to understand the significance of Web 2.0, Web 1.0 – the web prior to 2004 – must be defined.

Gopher sites predated early websites but both Gopher and websites prior to 2004 involved the publication of information by individuals and organizations and the consumption of information by other individuals and organizations. Anyone was free to publish information, they simply had to know the protocol and have the right tools. The early Internet was also the first popular method of addressing the long tail of information – specific topics, difficult to find through popular means because of their lack of a critical mass audience, published grassroots-style to small audiences of seekers.

Another popular use of Web 1.0 would have been e-mail. In fact, electronic communication predates the Internet going as far back as the late 1960s.[8] Prior to 2004 users were able to communicate electronically with one another for little cost or hassle. Multimedia was also published and consumed by people via the web prior to 2004: images in the form of ASCII art, later GIFs or JPEGs, audio in the form of MIDI files, and animation made possible by the animated GIF, Shockwave, and even early versions of Flash. Media types, other than text, were published to the early Internet, or Web 1.0, and people consumed them.

When examining Web 1.0, the web before 2004, it is clear that collaborative communication was taking place via e-mail. It is also clear that users were publishing content and other users were consuming it. It is clear that this content being published included subject matter that would have otherwise been difficult to access. And finally, that this content, while primarily in text form, also included a variety of media of various visual and auditory forms.

Post-2004 in the Web 2.0 world people are now e-mailing, MMSing, and instant messaging via the web. They are consuming Flash and Silverlight media, HD videos, 3D, and surround sound via the web. They are self-publishing via Facebook, Blogger, Twitter, and other social media. And people are personalizing with CSS, widgets, and social media.

When comparing Web 1.0 to Web 2.0, it is difficult to see the version change. Prior to 2004, people were using the web to collaborate, communicate, self-publish, access information, they were personalizing and consuming media. And post-2004 people are doing the same things. What has changed between then and now? Between Web 1.0 and Web 2.0? What happened in 2004? Other than the O'Reilly conference and the birth of a new slogan, not very much. People have always communicated, be it by carrier pigeon, letters via the postal service, e-mail, or instant message. Information has been stored, be it on scrolls, in books in library stacks, on microfiche, or on Gopher or websites on servers. The more information that is published and collected, the more difficult it is to find what one is looking for, so information has been organized. Be it by alphabet, by Dewey Decimal, by ISO standards like Dublin Core, or by social tagging. And people have published and consumed media, be it ASCII art and MIDI files, animated GIF, Flash, or HD-HTML5-3D-video. Prior to and since 2004, the activities are all the same: communication and being social, publishing and consuming information, organizing information, and being entertained.

In 2004 there was no new version release. There was no new web. Nothing happened that drastically changed the environment and people's behaviors. When taking two arbitrary points on an inclination, say point A as 1998, and comparing it to point B of 2004, the two points appear drastically different. While a great method to illustrate rate of change, this comparison ignores the fact that there has been constant, incremental change. Rather than a fundamental change or version in 2004, there has been an incremental and progressive evolution in technology and methodology over time that has facilitated what people were already doing. Technology and methodology evolve to better meet the expectations of users, and as this happens, the expectations of users increase, demanding more from the technology and methodology that are serving them. This is a constant cycle of progression and evolution.

What comes next and how to address it in establishing a strategy for social and mobile media means considering the Internet for what it is at its core: a network that provides communication and access to information. In addition, understanding the web as a domain of continuous evolution has been key in strategizing the Canadian Museum for Human Rights' future social and mobile media applications, and, in fact, its use of the web overall.

Ubiquity

In recognition of the constant evolution of the domain, and in anticipation that continued evolution will mean making more convenient the needs, practices, and behaviors that exist at present and have existed, the Canadian Museum for Human Rights is planning for absolute ubiquity.

The expectations of today's users are high. They are demanding and often impatient or unforgiving. In planning for an opening two years from the present, certain considerations have been highlighted. Youth, often considered early adopters of future popular trends (even if they proceed cautiously),[9] consume 11 linear hours of media per day, which is broken down as 7.5 hours spent multitasking, consuming more than one source at a time.[10] In the United States, the average person consumes 3.6 zettabytes (or the equivalency of approximately 5.1 trillion hard drives) worth of data per day.[11] Consider that 75% of teens have mobile phones yet decreasingly make phone calls.[12] Also consider that, in the autumn of 2009, Denmark and Finland both declared broadband a legal right, or that there has been a progression in a relatively short time from the modem, to cable, to Wi-Fi, and now to 4G MiFi and personal mobile hotspots. Access to information and the ability to communicate have become completely mobile and on-demand. Technology has become less visible through not only smaller devices but through blended technology such as the Social Media Garment by Electricfoxy that allows users to interact with social media sites through gesture,[13] or convergent technologies like eye-tracking, augmented reality, and social media brought together in Nokia's Mixed Reality Glasses.[14] Advances such as these that both respond to and encourage changes in expectations, that facilitate access to information and communication, must be considered and anticipated.

These current expressions of new blended technology and methodology among many others, and the understanding of the continued and constant evolution of the domain, provide insight helping to inform the CMHR's strategies for social and mobile media. There are innumerable other considerations that can be illustrated but the point is that the web, through constant incremental evolution, has been democratized to a point where certain behaviors and expectations are no longer those of the few, or growing minority, they are now and will be expectations and behaviors of the vast majority. These expectations call for support towards multi-tasked, on-demand service via a multitude of devices – whenever, wherever, however, thus the Museum's strategy for ubiquity.

An emerging strategy

In consideration of all the aforementioned insights and more, the approach from the CMHR will be to bring the web experience into the Museum itself and encourage multitasking for those who expect it, when and where appropriate. The strategy of the CMHR is to make the delineations of the pre-during-post museum web interaction scenario less easy to define; to encourage interactions between the Museum and visitors, visitors and visitors, and visitor types irrespective of time and place versus activity type. The Canadian Museum for Human Rights will not focus on the pre-during-post web scenario but rather on ubiquity and the consistent opportunity for access and interaction, using mobile and social media to facilitate these interactions.

Getting there

In order to build towards the goal of ubiquity, the Canadian Museum for Human Rights must start somewhere. This starting point is the data. Planning for ubiquity means acknowledging that the popular technology of today will be replaced with the technology of tomorrow. Not only is this an expectation, this is a key variable in the CMHR planning strategy. Building a data architecture than ensures the proper cataloging, preservation, administration, and delivery of data, according to standards is fundamental. Scalability, interoperability, and true efficiency are the main goals of the Museum's data architecture along with the principle of "store once, reference often". If the Museum can build its data architecture knowing that data only has to be created and stored once, then no matter what technology or device becomes popular later, the content will not need to be recreated, only the delivery protocol will require addressing. And even then, only if subject to a new delivery standard.

Stored once, data is then pushed or pulled from the store to whatever device requests it. Context is held or delivered separately. The device then contextualizes the fed content as appropriate. A clear and strict separation between content and presentation is crucial. This will enable the changing of the device without affecting the content, and or the editing of the content (or addition and growth of the content) without the need to change the presentation or the end device. This is the very high-level basis for the CMHR's data architecture. Accepting Museum data, user-created data, partner or shared data, and media of all kinds and from all origins, such a basic approach will set the foundation for efficient flexibility. Effort can then be spent on actually dialoguing or interacting with the audience through their preferred means no matter how these means evolve.

Imagine

Imagine a photo of a hungry child sitting against a stone wall and the accompanying text describing the *Right to Food*. The photo and text exist in the CMHR data structure and could simultaneously be viewed by any of the following scenarios, each occasion wrapped in the appropriate context (including sizing the photo, abbreviating the text, etc.):

- two visitors discuss the image as it appears on a digital kiosk in the gallery space next to a series of moral questions;
- a researcher sitting at her desk using the Archive Library Collection Management (ALM) system digitally shares a reference of the image and all associated metadata with a colleague as they co-author a paper;
- a university student visiting the museum reference center prints out the image in high resolution along with the list of where it has been used and all user related comments;
- someone sitting in his home rates this photo from the CMHR website's media gallery;
- a person using his smartphone sitting in a piazza across the globe views the photo against its original backdrop via augmented reality;
- a student adding music to images via a collaborative web project hosted by Amnesty International includes this image in her slide show;
- a teacher projects the photo to the class as part of a digital learning resource;
- a teenager logs in to Facebook and sees the image featured in ArtShare, "likes" it, and the image and associated info are delivered to the Facebook pages of this person's "friends";
- a student tagging a collection of photos on Flickr Commons from their tablet makes this photo part of

her personal Flickr group.

An infinite amount of uses could be described that all involve the use of this one image and its accompanying text and data as part of a visitor's social interaction with both the Museum and with other visitors. Over time, as the domain evolves incrementally, the CMHR in its desire to meet the expectations of its audience, will have to evolve as well. Without the appropriate architecture, the Museum would not be able to offer these opportunities in an efficient way.

The Canadian Museum for Human Rights is only just beginning to operate. The building is being constructed, exhibition and museum program designs are underway, staff is being hired, processes and procedures established, and many infrastructures architected. The canvas is clean. This is a unique opportunity for those involved in the Museum's creation and growth, and the foundations are being set. The dialogue between the Museum and its audience has begun. If established properly, both in methodology and structure, the Museum's continued opportunity for dialogue and interaction will result in an ongoing reciprocal relationship where the museum informs its visitors, and its visitors inform the museum. Social and mobile media can be great facilitators of this interactivity. They will be used by the CMHR in a way that strategically addresses the evolution of behavior and expectations, matching the nature of the web domain itself.

Notes

1. Forks North Portage Corporation, http://www.theforks.com/history.

2. Office of the Prime Minister of Canada, http://pm.gc.ca/eng/media.asp?id=2355.

3. Lord, Gail and Lord, Barry; Manual of Museum Exhibitions (2002) p.469.

4. Nathan Shedroff, Experience Design (2001).

5. Filippini-Fantoni, Silvia; Bowen, Jonathen; Bookmarking in Museums: Extending the Museum Experience Beyond the Visit?, Archives and Museum Informatics: Museums and the Web (2007)

Granger, Lisa M; The Visitor Has Left the Building: Using Technology to Extend the Museum Experience, John F Kennedy University, (2005)

Granger, Lisa; Tech Tags: Extending the Visitor Experience at the Tech Museum of Innovation, Dimensions, (July 2006)

Haley Goldman, Kate; Cell Phones and Exhibitions 2.0: Moving Beyond the Pilot Stage, Archives and Museum Informatics: Museums and the Web (2007)

MuseumPods Inc., Museum QR Mobile Podcast Media Directory, 2005-2010, http://www.museumpods.com/

Richardson, Jim; Co-Producing the Museum, Sumo (2009) http://www.museummarketing.co.uk/2009/12/07/co-producing-the-museum/

6. http://oreilly.com/web2/archive/what-is-web-20.html.

7. There are many software or hardware development methodologies yet this is a commonality (release and versioning) among them all.

8. http://www.nethistory.info/History%20of%20the%20Internet/email.html.

9. Amanda Lenhart, Kristen Purcell, Aaron Smith, Kathryn Zickuhr, Social Media and Young Adults, Pew (February 2010).

10. Victoria J. Rideout, M.A., Ulla G. Foehr, Ph.D., Donald F. Roberts, Ph.D.Generation; M2: Media in the Lives of 8-18 Year Olds, Kaiser Family Study (January 2010).

11. Bohn, Roger E; Short, James E; How Much Information? 2009 Report on American Consumers, (January 2010).

12. Amanda Lenhart, Pew Research Center's Internet & American Life ProjectRich Ling, ITU, Copenhagen, Telenor and University of Michigan Scott Campbell, University of MichiganKristen Purcell, Pew Research Center's Internet & American Life Project, Teens and Mobile Phones, (April 2010).

13. http://www.electricfoxy.com/ping/.

14.http://www.youtube.com/watch?v=CGwvZWyLiBU

http://research.nokia.com/files/NTI_MARA_-_June_2009.pdf

Acknowledgement

The author wishes to acknowledge Victoria Dickenson and Michael Mouw as two main collaborators in daily schemes, ideas, brainstorms, big dreams, and general nonsense.

Virtual Worlds as Technologies for Lifelong Learning and the Implications for Museums

JOE SANCHEZ

Rutgers University

Online, three-dimensional virtual world environments are becoming more widely-used knowledge and social-interaction tools, and will become another part of the electronic sociotechnical system that people use for communication and lifelong learning in the foreseeable future. As museums accept the challenge of community engagement and the rising demand for online access to exhibits, virtual environments provide an alternate venue for extending their geographical reach while still providing an immersive experience for visitors. Like the internet, online virtual world environments such as Second Life® are already being used for numerous kinds of human activity, and they provide both synchronous and asynchronous capabilities for everyday interaction. However, in contrast with text-based online learning settings and web-based computer-supported collaboration tools, 3D virtual world environments create a sense of physical place and presence through the visual projection of oneself and other individuals in the form of avatars interacting within three-dimensional physical spaces; for example, a virtual neighborhood, a virtual classroom, a virtual museum, and a virtual rainforest are all sites for lifelong learning activity.

This paper presents an approach that lifelong learning researchers, technologists, educators, and students can consider for their entry into 3D virtual world technology and virtual learning environments. I will discuss ways that museums, as users of the technology, can enhance their collaboration with colleagues and can extend some very practical research and educational operations for museums into virtual spaces. I will draw on findings from a multi-year, ongoing research project conducted in Second Life and discuss the Educators Coop, the first long-term residential virtual community of educators and researchers in Second Life (SL). The informational design and infrastructure of the Educators Coop together serve as an emerging model for community-based museum environments targeting lifelong learning.

Online virtual worlds are emerging technologies, and thus I will provide a brief review of the literature that examines the apparent distinctions between virtual world and real world learning, and how researchers are negotiating that tension. Then I will move to the primary focus of the paper: the useful applications ascertained from the ongoing Educators Coop model and the numerous education professionals with whom the group is collaborating, notwithstanding some inevitable and interesting challenges. Finally I will present future directions for museums interested in lifelong learning opportunities in virtual worlds.

Virtual and real learning: breaking down the distinctions

The apparent distinctions between virtual world learning and real world learning are problematic in the literature, and researchers are negotiating new descriptions of learners' experiences. According to de Nood and Attema (2006) in open-ended online simulations, the:

> *... distinction between the physical (real) world and the virtual world tends to disappear. As the distinction between these two*

> *worlds fades in the experience of the visitors, one speaks of "inter-reality." When these virtual worlds continue to grow and develop, they would then seem to offer practically unexplored opportunities for our society, in economic, cultural and social terms. (p. 3)*

Indeed, Steinkuehler and Williams refer to online games and simulations and the social networks they can generate as "third places" (2006). Bloomfield emphasizes a key feature in his definition of virtual spaces and, for example, he argues that "Second Life represents a persistent virtual world because the servers that control them run continuously, allowing software programs to alter the world even when no members are present, and the worlds will continue as long as the game publishers choose to sustain them" (2007, p. 18). These third places continue to persist in time, and online learners can revisit earlier locations and simulations without the virtual environment resetting itself or "starting over." This feature of persistence allows for the archiving and growth of new knowledge places and structures.

From a different perspective, Ursula Franklin suggests that approaching games and simulations as tools or technologies helps us widen our view and see technologies – for example, online virtual simulations – as systems of social practice (Franklin, 1999). For educational applications, this online simulation environment is another spatial dimension (albeit virtual) into which participants extend themselves while maintaining a sense of realness or reality. That is, one experiences being co-present with others in a kind of continuum that includes both the virtual and the real, and this experience can make the distinction essentially irrelevant.

Users of SL include a growing network of connected real life individuals and groups collaborating virtually in global educational and even social change activity. Drawing on tacit behavioral patterns from their life and work environments, these *homo virtualis* users (Jarmon, 2009) are improvising new forms of social interaction and are discovering new ways of making sense of and talking about their embodied social experiences. Based on evidence provided by these users, it is becoming increasingly evident that ongoing iterations of new, repeated, and improvised experiences begin to dissolve some formerly solid distinctions between the experience of the "virtual" and the experience of the "real" (Jarmon, 2009). Therefore, for the purposes of this paper, I consider lifelong learners' actions and intentions, regardless of "where" they may be occurring.

Virtual worlds for lifelong learning: The Educators Coop Model

In this section, I present the primary focus of the paper, the Educators Coop model of a learning community, and I will discuss useful applications ascertained from its ongoing development and from the numerous education professionals and researchers who are collaborating within the group. First, I will describe how my observations are grounded in concrete data as part of a long-term research study. Second,

I will consider how the design of a virtual space in SL can deliberately support collaboration and the growth of a community culture. Third, I will take an in-depth look at Wenger's ideas about communities of practice and at how including four of Wenger's key components in the design of a lifelong learning community can impact the experiential learning of its members. Fourth, I will present concrete examples that illustrate how the Educators Coop is evolving over time as a learning community. Finally, I will end this section with brief perspectives on high school girls who created museum exhibits focusing on science, math, and technology in a virtual summer camp.

IRB-approved long-term research study

The observations presented in this section are grounded in concrete data as part of long-term research study. This ongoing IRB-approved research project being conducted in SL focuses on collaboration and interactions in virtual worlds, specifically among the members of the Educators Coop. The Educators Coop (signifying both co-op and a space of close proximity) is a virtual non-profit residential community of university faculty, librarians, and K-12 teachers actively teaching or conducting research in SL. Participants volunteered to join the community and are from 32 different educational institutions. The majority first met in SL and have only known one another virtually. They meet regularly in SL to share virtual world teaching strategies, to design virtual world and real world research projects, to collaborate on interdisciplinary conference sessions, and most importantly, to create a support system for geospatially separated academic practitioners interested in teaching and conducting research in virtual worlds (Jarmon & Sanchez, 2009; 2008a; 2008b). Preliminary results indicate that these researchers and educators are using the virtual world in very practical and concrete ways to carry out their work.

During the last 26 months, I have collected hundreds of chat logs, conducted five online surveys, held four virtual focus group sessions, conducted 46 in-depth virtual interviews in SL with members of the Educators Coop, and held a community retreat in SL. The interviewees ranged in length of residency in SL from one month to over a year at the time of the interviews.

Collaboration and community: designing a virtual space

The design of a virtual space can support the kinds of collaboration that may occur there and lead to the growth of a learning community. In this section, I discuss how the Educators Coop has been designed. Museums entering a virtual space should not only consider the design space of their exhibits, but they will also want to consider how design impacts the user experience and how exhibits can be designed with collaborative spaces intertwined. The mission of the Educators Coop is to create a virtual community of educators and researchers interested in learning how to teach and conduct research in a virtual world environment. The community was designed based partly on Wenger's model of communities of practice (CoP) (Wenger, 1998), and Wenger et al. (2002) suggest the three elements of a CoP: the domain, the community, and the practice.

The design of the domain or the shared identity held by members of the Educators Coop is a shared interest in teaching and research in SL. All members that join the group must explicitly state their teaching and research goals in their application for membership. The design of the community allows members to participate and initiate joint activities. It is an inclusive environment that allows individual members to make contributions and to openly negotiate the unfolding structure of interactions and experiences. Members are encouraged to interact and build relationships with other members so that they can learn from each other. The open design of the practice involves the members of the Educators Coop as practitioners, as educators in the general sense but specifically as professors, librarians, K-12 teachers, or researchers who thus share a baseline of common knowledge and goals.

A key concept of communities of practice is situated learning, or the relationship between learning and the social situations within which it occurs (Lave & Wenger, 1991). Unlike a traditional classroom model of learning, situated learning places the learner within an integral system that includes relationships (newcomers and old-timers) and authentic environments (the external world). Lave and Wenger use the example of a tailor and an apprentice when describing situated learning wherein the apprentice learns basic skills, begins to work on more complex tasks, and eventually learns to more advanced skills. With time and through practice, the apprentice learns the process of making garments and the tasks involved at each stage. Lave and Wenger call this process legitimate peripheral participation, while the act of learning through an authentic field-based environment is referred to as situated learning.

The Educators Coop is designed in a way to support both situated learning and legitimate peripheral participation. For example, the community was designed to have all learning and interaction occur within the virtual world environment. A large community sandbox was constructed so that members could develop virtual world building skills while situated within the island, and they are often focused on building their virtual homes or presentation and meeting areas for their students, all situated learning activities. Because members of the group have various experience levels with the SL platform and its multiple tools, the members with more time or expertise in SL often provide guidance to newcomers.

Collaboration and the overall user experience within SL can be viewed as a socio-technical system of developers, users, groups of users and the software (Sanchez, 2007). Since a goal of the Educators Coop is to provide a place for collaboration for geospatially separated educators, the design of the model allows for members to use the virtual world as a work tool. The organization of work in the community has become a system of people interacting with one another and with the technologies available to them. Lucy Suchman (1987) has referred to this kind of organized work activity as situated action, and she argues that "work activities in every case take place at particular times, in particular places, and in relation to specific social and technological circumstances" (p. 50). So the design of the Educators Coop model intentionally included elements that, it was hoped, would be conducive to collaboration and the emergence of a learning community.

Designing an experiential learning community

Now I take a closer look at Wenger's ideas about communities of practice and at how including four of Wenger's key components in the design of a lifelong learning community can impact the experiential learning and growing sense of community among its members. Engagement with virtual worlds in a learning setting can contribute to the facilitation of life-long learning that extends beyond the confines of the classroom, can have the potential to generate feelings of copresence and connection among participants in and outside of virtual worlds, and can have the potential to enrich the lives of learners of all ages (Jarmon et al., 2009, in press). In using Wenger's notion of a learning community, a special kind of community of practice (Wenger, 1998), I specifically focus on four characteristics of these communities: their shared repertoire, mutual engagement, joint enterprises, and how they negotiate meaning and practice (p. 72). In this section I will examine how the Educators Coop model integrates these four community features into its design.

In recent research on experiential learning and situated

learning spaces, Kolb and Kolb (2009) update their earlier theoretical framework and remind us that their experiential learning model includes "socialization into a wider community of practice that involves membership, identity formation, [and] transitioning from novice to expert" (pp. 24-25) – not unlike Wenger's framework. However, time and immersion are required by members to discover and to better understand the characteristics of new virtual learning affordances in SL. Following a design that integrated these four features into its structure, the community members use their shared repertoires, mutual engagement, joint enterprises, and continuous negotiation of meaning and practice to operationalize the model of the Educators Coop. Next, I take a more in-depth look at each of these four features.

Shared repertoire: Two elements of the Educators Coop model explicitly create the conditions for its members to have a recognizably shared repertoire that draws on or creates common ground for those who at first are largely complete strangers: eligibility criteria for membership and the five-step application procedure.

The eligibility and selection criteria for membership are simple and straightforward, but they are also relatively narrow and ensure that accepted applicants to the community have some basic background, experiences, and aspirations in common. The most critical criterion is that the applicant must be an academic educator or an educational researcher, and this category includes academic faculty, graduate student instructors, K-12 teachers, and librarians. Consultants, professional development trainers, and technical or information technology coordinators do not typically qualify. In this way, all members accepted to the community know immediately that the other members they meet in the Educators Coop in SL share general, if not passionate, interest and experience in education. As educators, they can assume a basic shared repertoire of some educational experience, some teaching skills, and an interest in learning.

The second element of the model that explicitly creates some common ground for applicants in the community is the five-step application procedure; again, all members accepted to the community know that everyone else has gone through this process and so it becomes part of their shared repertoire. The steps are as follows:

- Step 1 – Eligible applicants email to us their avatar name, real name, email address, a web site listing their academic affiliation or presence, and standard contact information. In the same e-mail, eligible applicants write two paragraphs describing (a) what they are interested in and imagine doing in SL as an educator/researcher, and (b) what they are interested in teaching (topics, workshops) and/or conducting research about (topics) in collaboration with other residents.
- Step 2 – A brief telephone, Skype, or SL-voiced interview is conducted to verify who the applicants are and to address any of their questions about the

joining the Educators Coop.

- Step 3 – Using emailed documents, applicants must agree to and sign the IRB Research Consent form, the modified SL Terms of Service form, and a form describing the general Educators Coop civil rules.
- Step 4 – Since the Educators Coop in SL is a residential community, the applicants select their parcel(s) of land on the community's virtual islands and submit their land requests through email. Unlike most simulations offering virtual land in SL, as a non-profit, the Educators Coop model is designed to cover only its costs in administering the community islands (no one draws a salary) and so it can offer applicants the lowest possible annual leasing rates allowed by Linden Lab land costs. Invoicing is conducted through email.
- Step 5 – When payment of the lease is provided, the settings on the applicants' virtual parcels are adjusted to indicate them as the new owners with full building permissions and other capabilities. Applicants are then invited to join the special members-only SL group for the community and begin designing and creating their presence in the community.

These five steps ensure that all members are *bona fide* educators, have articulated an initial vision of their activities in the community, are participating in the ongoing research project, have agreed to the same terms of service and special community civil rules, have made at least a one-year commitment to the residential community, and have joined the special SL group. All these shared actions and the intentionality embedded within them contribute to the members' sense of a basic shared repertoire as well as a sense of community identity.

When designing a virtual museum, administrators must realize that the virtual museum cannot be all things to all people. Specific exhibits should be designed to target communities of users with a shared repertoire. As in the real museum space, teens will have different needs and interests than adults. Exhibits should be created to target specific users so that the likelihood of an ongoing interest group can form.

Mutual engagement: The second feature integrated into the design of the model is mutual engagement. That is, specific elements of the e-infrastructure of the Educators Coop were intentionally designed to support the mutual engagement and collaboration activities of the residential community members. Launched in July 2007, the goal of the Educators Coop is to provide educators and researchers with a unique residential environment from which to begin exploring, collaborating, teaching, learning, and conducting research in SL. Residential refers to the design of the virtual island, because it is designed to look like a neighborhood and each resident has one or more parcels of land where they build a private virtual residence or whatever they create as their own presence in the community (some have made gardens). In the vast virtual environment of SL, a user can inadvertently cross paths with people engaged in essentially the full range of human activities, the good and the bad. Importantly, however, proximity to other educators was deliberately designed into the organizational and virtual infrastructure of the Educators Coop. All residents are informed participants in the research project and understand that the purpose of the research project is to analyze cross-disciplinary relationships and collaborations over time.

The architecture of the Educators Coop was designed to encourage mutual engagement, support social exchange, facilitate navigation across the island, provide easy access to community information, and to support group collaboration. Initially, the Educators Coop was comprised of sixteen virtual acres of land divided into a central mesa and sixty 1024-square-meter parcels for the residents' homes. Each quadrant of the island is marked with a large colored flag (red, yellow, blue, and green), and at the base of each flag is a small gathering space featuring benches and landscaping for that neighborhood.

Prior to the residents occupying their parcels, a sense of place was intentionally created by building a clearly visual navigation system (sidewalks), social spaces (neighborhood gathering points), and an information hub (the Mesa), all surrounded by trees and landscaping. Originally, a six-meter-wide waterway surrounded the island and connected it to neighboring islands in the Library Alliance archipelago, a group of connected islands administered by the Alliance

Library System based in East Peoria, Illinois (USA).

The Mesa, a key e-infrastructure support of mutual engagement among the members, is the center of the island and it features a community sandbox, a large group discussion area, a community calendar, a bulletin board, and a small group discussion area. The sandbox is a collaborative building space where residents can create and test objects. As indicated earlier, the land where users are allowed to experiment and build is relatively expensive in SL, and the sandbox is a valuable community commodity for residents.

When first launched in 2007, residents joining the Educators Coop had the option to choose from a variety of seven house designs to place on their land or to build their own structure. The seven house designs were selected because they fit within a parcel of land and so allowed for the intentional proximity of members; they were designed well for avatar movement and matched the overall aesthetic of the residential environment. These elements combined to contribute initially to a unique sense of community, identity, and mutual engagement among the members.

For museums, allowing virtual guests to create exhibits or participate in the design of their community will allow users to become mutually engaged in the virtual space. Engagement in the virtual space will create opportunities for collaboration and a feeling of ownership in the museum that is not as easily attainable outside a virtual space. The idea of museum patrons shifting from visitors or guests of a museum to residents or co-creators of a museum is possible by following a model of interaction similar to the Educators Coop.

Joint enterprise using SL communication tools: The third of the four features integrated into the design of the model to help create an experiential learning community is joint enterprise. In other words, the model makes explicit use of certain communication affordances built into the SL virtual platform that can make it relatively easy for community members to embark on joint activities.

Research (Jarmon & Sanchez, 2009) at the Educators Coop has helped to identify an initial typology of text and voice communication practices in SL (Fig. 1). There are multiple modalities within which interactions and joint enterprises can be undertaken, and several different communication tools may be in use at the same time. That is, in the virtual world it is possible to participate in more than one interaction at a time, making the nature of participation in joint enterprises different than in non-virtual learning environments. Communication in SL can occur in text chat and through voice chat using a headset. Chat can also be through an open public channel, a private instant messaging channel, or through a group messaging channel. During events such as meetings or conferences, communication may occur through streaming media into SL from a real life (RL) place. Thus, a group interaction – a joint enterprise – in SL may involve a variety of public text chat mixed with private instant messaging, voice, and announcements sent and received through group chat. Fig. 1 identifies ways in

Text and voice communication at the Educators Coop in Second Life		
Interaction in SL	**Interaction occurs among**	**Examples**
Public chat Public voice	Two or more interactants who are virtually co-present; "overhearing" possible at 20 meters	interacting avatars chat in a public area about teaching practices in SL
Private messaging Private voice/call	Two interactants; may or may not be virtually co-present	One participant asks another for her location in SL
Group chat (only members of the group have access to the communication)	Multiple members of the group; may or may not be virtually co-present	A group member announces that an event is about to begin; members of a group discuss a guest speaker's lecture, adding information, assessments
Group notice	Group officers to group members	Group owner sends a notice of an upcoming event to its members
Group proposal	Group officers to group members	Group members asked to vote on a resolution
Streaming audio	One (real life broadcaster) to many (virtually co-present recipients); typically, recipients can participate in additional interaction via chat messaging or using voice chat	Video is streamed from a "real life" conference into SL and all participate in Q&A

Figure 1: An initial typology of text and voice communication practices observed at the Educators Coop in SL (adapted from Jarmon & Sanchez, 2009).

which text and voice communication may occur among two or more community members in SL, groups of community members, or among virtually co-present members and real life face-to-face members.

All these forms of communication afforded by the SL platform are used and encouraged in the Educators Coop model. The ability to communicate in so many different ways has led to numerous joint interactions among members, and has helped to create bonds within the whole community. The interactions and joint enterprises are part of the experiential learning and discovery process among members.

Negotiating meaning and practice: collaboration: The fourth and last feature integrated into the design of the Educators Coop model to help create an experiential learning community is negotiating meaning and practice. In other words, the initial design intentionally left open much of the organizational and social structure to be improvised and determined by the members themselves over time; this has led to the emergence of collaboration among members as they struggle together to become a community and make sense of what that community means to them.

Collaboration can occur because SL virtual technology provides conditions for an experiential, embodied, and social reality, and this social reality provides a virtual "new space" wherein existing communication practices and social networking tools are converging. As Suchman has observed, "a frame of artful integration emphasizes the ways in which new things are made up out of reconfigurations and extensions to familiar environments and forms of action" (2002, p. 144). Members in the Educators Coop are improvising from prior patterns of behavior newly reinterpreted in virtual social spaces that are filled with new tools, and members begin to perform differently. As new practices are negotiated throughout SL, users connect, build relationships (Nardi, 2005), pursue objectives, and act with embodied purpose; in some cases, communities of practice are forming or reestablishing themselves in virtual spaces. These generative and performative capabilities extending into virtual worlds are experiential by nature and lead to new practices among members.

To conclude this subsection, community members are using their shared repertoires, mutual engagement, joint enterprises, and their continuous negotiation of meanings and practices; they are essentially operationalizing these four features of the Educators Coop model design, thereby creating their own experiential learning community over time. A virtual museum designed with the ideas of Wenger's CoP model may have similar results within its community of users.

Evolving over time as a learning community: examples

To illustrate some of the ideas described above, I now turn to the fourth part of the primary focus of the paper (the Educators Coop model of a learning community) to discuss concrete examples of how the experiential learning community is evolving over time. The members' activities may have roots in academic collaborations, but the lifelong learning aspects of their various long-term projects will become evident as they become involved in behaviors and efforts beyond the purely academic.

I begin with an example of how the model serves as a rapid collaboration platform (Jarmon & Sanchez, 2008a). According to the interviewees, collaboration is one of the most significant ways they use the Educators Coop, and that collaboration takes many forms. Conference panels, papers, presentations, course projects, and mentoring are some examples of the types of collaboration reported. Members asserted that the Educators Coop presented many unexpected opportunities for collaboration on real-world efforts in the design of assistive technologies for people with special needs, in creating poetry and music, and in working with real-world non-profits. Several members have become co-presenters in real-world academic conferences even though they have never met in person; their first connection and their collaboration in planning occurred through the Educators Coop in SL.

Typical of the members' observations is this: "On Educators Coop island, there are at least three people now interested in assistive technology/special education." Two of these members have continued to focus on accessibility in virtual worlds and have presented at conferences; they are collaborating with the larger network of people in SL working on accessibility issues – academic, nonacademic, non-profits, and people living with disabilities – and these members have created an Accessibility Island in SL.

Several examples of collaboration initiated through the Educators Coop are also examples of the cyclical nature of lifelong learning and the novice-expert feature of an evolving learning community. In the second example, an assistant professor in library and information studies adopted a mentoring role with a doctoral student and fellow member of the Educators Coop from a different university. They had known each other in SL for only three months and had never met prior to their membership in the Educators Coop. During an informal chat on the Mesa, they began discussing an upcoming conference and realized that each was planning to submit a paper about teaching in SL when the next call for proposals was issued for the following year's event (the deadline for submitting to the next conference had already passed). However, the assistant professor mentioned that she had a panel session accepted for the upcoming conference occurring in three months. If the doctoral student was interested, she offered to ask the other panelists if they would mind adding another speaker. The student agreed; three months later, he and the assistant professor met in person for the first time, a few minutes before the panel session began, at the Association for Library and Information Science Education conference.

Thus, lifelong learning can be a cyclical process where learners participate in an evolving process from student to teacher, master to apprentice, and back again. In our example here, the assistant professor is still in a learning role as she works toward tenure; yet, she plays another role as a mentor and expert when she creates and offers an opportunity for the novice student to serve as a panelist at a professional conference. Furthermore, the virtual world environment can allow for the levelling of some preexisting social-hierarchical boundaries and status differences within this emerging community of practice. Virtual worlds, and particularly the Educators Coop model of an evolving learning community, can provide a supportive environment for the continuous learning cycle and the lifelong learning process.

It is important to note that within the model, members have already completed standard academic degrees; they are continuing their learning and discovery of new knowledge that often extends beyond the narrower focus of their academic disciplines and leads to lifelong learning opportunities. For example, the e-infrastructure of SL provides a mechanism for users to join up to 25 groups in the virtual world, and to create new groups for a very small fee ($100 Lindens; about 35 cents US). Members in the Educators Coop have affiliated with a wide array of groups unrelated to their academic disciplines and reflecting extended engagement in other nonacademic and informal learning activities. Fig. 2 offers a glimpse of the group affiliations of some community members.

The diverse participants in the Educators Coop community, their students, and their guests, create what Gee has called "cross-functional affiliations" (2007, p. 327). They may have their "real" and "virtual" areas of specialization, but

Member's academic discipline	Groups joined in SL (not related to academic discipine)
Information Science	Music Academy Group; Study of Religion in Virtual Worlds
Library Science	Friends of The Botanical Gardens
Organizational Communication	Tube2SL Actors Guild (SL movie production/video broadcasting company
Educational Technology	ZeroG SkyDancers; End Homelessness; Games for Change; The D.I.D.I. Initiative (supports youth launching
Visual Cognition	Rainbow Gardens Pet Memorial Group
Nursing	Diegolanders (San Diego fans); Moms of Forest House (for moms); Studio Wikitecture (virtual architecture)
Special Education	NeuroMuscular Junction Wellness Center; Style Enabled Modeling Agency
Anthropology	Cognitive Science; Nymf's Science Circle
Educational Technology	MAD Agency Fashion & Design; Science Fiction Rangers
Communication	SNAPSHOTERS – SL Photography Group; Storytelling Guild of SL
Education	AZTECA Mexican Embassy; Botanica Garden Club; Lions Club International; Virtual Helping Hands Group
Geography	Bliss Gardens
Library Science	Japanese Cultural Group
Information Science	Roleplay Collective

Figure 2: Selected Educators Coop members' academic disciplines and groups joined in SL (not related to their academic discipline).

they also collaborate and share their knowledge and resources across what may have been perceived as social boundaries in their face-to-face lives. In the evolving Educators Coop learning community model, all these participants can be seen as learners with expanded roles: learners as scientists, learners as designers, learners as researchers, learners as communicators, learners as collaborators. The ethic in this learning community model is generosity: passing it on to colleagues and peers and thereby widely extending participants' own support networks.

With these examples, one can begin to glimpse the robust learning environment of not only the online virtual world platform of SL but also of the design features of the Educators Coop model that have led to the ongoing evolution of this learning community.

Lifelong learning

As I've demonstrated, learning opportunities extend beyond formal academic environments in SL, and I end this section with some examples from other perspectives on using SL for lifelong learning. For example, representatives of approximately 80 non-profit organizations from around the world have formed a virtual community in SL called the Non-Profit Commons: members hold weekly virtual meetings, help one another create virtual spaces, convene conferences, hold fundraisers, and raise awareness about their causes. Members are learning from one another on a continuous basis, and the SL infrastructure allows for this new kind of learning collaboration.

Girlstart, an educational non-profit organization in Austin, Texas (USA), sponsored an Information Technology Summer Academy in Teen Second Life during the summer of 2008 for 40 high school girls and their eight counsellors. In addition to its standard STEM literacy activity, the Girlstart camp curriculum included sessions on gaming, career pathways, social change, and gender equality. Members of the Educators Coop community designed and led the summer camp and its learning activities in SL, and rapid collaboration marked the campers' experiences. As one camper noted about learning in the virtual world, "I liked how easy it was to interact with others."

Another recent study examined how SL may be perceived and understood by older adults in a volunteer social seminar setting for retirees (Jarmon, et al., 2009). Preliminary findings indicate that engagement with virtual worlds in a course setting:

1. can contribute to the facilitation of lifelong learning that extends beyond the confines of the classroom
2. has the potential to generate feelings of copresence and connection among participants in and outside of virtual worlds;
3. can provide a context for considering how new technologies have the potential to enrich the lives of older adults.

The study was conducted during a six-session introductory

course on SL offered by a member of the Educators Coop to older adults participating in the Osher Lifelong Learning Institute (OLLI) program at a large southwestern research university. The OLLI program provides intellectual enrichment for mostly retired individuals and is conducted in a structured academic setting. There were no grades or credit assigned for this course, and the seminar presented a wide-ranging introduction to SL using virtual guest speakers and virtual field trips. The results contribute to our understanding of the potential for SL to provide a framework for establishing virtual lifelong learning environments and expanding the access of older adults to extended social connections.

Using online virtual worlds such as SL can have critical implications for some special needs populations as well as for older adults. For example, when stroke victims visited the protected virtual area in SL for people with disabilities called SL Dreams, they reported that the experience of seeing themselves walking aided in their recovery (Stein, 2007). In addition, people with autistic disorders use SL for therapeutic purposes (Mangan, 2008; Au, 2008). However, the virtual infrastructure of SL also allows for several people to be "embodied" into a single avatar. This is the case with the avatar known as Wilde Cunningham, who represents nine people with cerebral palsy. This example raises interesting questions about social interaction involving a virtual participant who is representing multiple actual persons; a new level of complexity adds to the idea of co-present participants and access to opportunities for learning in virtual personal learning environments:

> *Wilde Cunningham is an avatar controlled by a group of nine adults with cerebral palsy (and their nurse) at the day-care program they attend in Massachusetts. The group members are aged 30 to 70 and comprise four men and five women. Most of them are wheelchair users and rely on their caretakers for almost all aspects of their daily lives. Yet in Second Life they have built their own houses, have pets, gardens, even a baseball field. They also have many close friends and a large social network. "Second Life gives me the chance to be the person I feel I was born to be," says John S, 32, one of the group. "Being in Second Life is how I imagine an innocent man who had been locked up wrongly feels when he is finally set free. In Second Life I get to call the shots." For John S, the virtual world is all about being free from his disability but for Simon Stevens, who also suffers from cerebral palsy, it is equally about making disabled people visible. (Deeley, 2007).*

In the short-term, user acceptance of 3D virtual world environments will continue to be an important challenge to overcome (Fetscherin & Lattemann, 2007, p. 20). That there is no predefined structure on how to use virtual worlds suggests how critical innovative instructional design can be to facilitate learning in such environments (Jarmon et al., 2009, in press; Mayrath et al., 2007; Sanchez, 2007). Lynch and Tunstall (2008) have suggested that projects attempting to develop educational simulation games require a development framework and design process that is integrated into course design; is engaging, relevant, and useful for students; and is flexible, cost effective, and reusable (p. 383). More basic research is needed to demonstrate both the effective design of learning activities in virtual spaces and how virtual spaces are influencing what we imagine education itself will become, regardless of age group.

In this section, I have discussed in some detail the Educators Coop model of a learning community, and I have shown how the model's design features can deliberately support collaboration and the growth of an experiential, lifelong learning community that can evolve over time. A number of examples were presented to illustrate those design features and to show how 3D online virtual world environments can provide lifelong learning opportunities that can extend beyond formal academic environments to include learners of all ages and those with special needs.

Future research directions

Two major initiatives let us see the likelihood of how future trends in the adoption of online virtual world technologies for lifelong learning may unfold. First, in the case of the global corporation, IBM, we must take into account that as of the publication date of this volume, over 6,000 IBM employees go to work every day in a secure SL environment. IBM has already held a Virtual World Conference and an Annual Meeting in SL, and IBM estimates that the conference was held at one-fifth its real world cost (Linden Lab, 2009).

Organizations will need to continually improve and educate their workforce to increase capacity for knowledge creation; the use of shared virtual environments and virtual worlds such as SL is an innovative way to share information and create opportunities for lifelong continuous learning within and across organizations. More research is needed to create metrics for assessing the effectiveness of different virtual models under diverse conditions.

Community will continue to increase in importance and research is needed to define the scope of these new human/virtual learning experiences. New epistemologies are likely to arise as our ways of knowing and learning increasingly become a blend of the traditional and the virtual. Instruction in information literacy is likely to become more critical for workforces as well as for societies; the skills involved may include not only discerning and sorting information, but also collaborating in its exploration, discovery, and creation in virtual environments.

Second, the viable Educators Coop model is robust with its unique informational, community, and infrastructure design, and our hope is that it can serve as an emerging model for next generation environments for lifelong learning. For example, the University of Texas System is using the Educators Coop as its model for an evolving learning community, extending its educational operations into SL and reducing overall costs in a statewide project with 16 campuses that host almost 200,000 students and 7,500 faculty. Pioneering cyber-learning and cyber-infrastructure, the UT System virtual initiative is the first effort in history envisioned at such a large scale. Designed for extensive collaboration with educators and educational institutions globally, the project is networked, customizable, and computationally and socially rich to ensure wide adaptability and expansion in diverse educational environments. Grounded in educational principles and integrating research and assessment, the initiative is based on the most current evidence-based research on pedagogy in virtual learning spaces.

In an ecology of virtual contexts, learners inhabit a broader landscape of their own and others' making that allows them to be teachers, designers, researchers, communicators, and collaborators. Active engagement in the virtual world will pull us into a broader landscape where "real life" plus "virtual life" yields an interreality that is rapidly growing vaster as it is simultaneously being created by and being inhabited by homo virtualis (Jarmon, 2009)

Conclusion

Online virtual world environments provide a new avenue for museums to reach out to their patrons. In this paper, I have presented the informational design and infrastructure of the Educators Coop that I hope can serve as an emerging model for next generation museum exhibits for lifelong learning. I have examined some of the tensions researchers are negotiating in terms of the apparent distinctions between virtual world and real world learning, and I have looked briefly at the emergence of a virtual world learning community over time. I have drawn on findings from an ongoing research project conducted at the Educators Coop in SL, and have provided concrete examples of various ways that these learning community members, as users of the technology, have enhanced their collaboration with colleagues by using the extensive affordances of the virtual world as a tool for knowledge and social interaction. This approach to lifelong learning, I hope, will be considered by museum designers and educators as they debate their entry into 3D virtual world environments.

As networked virtual worlds become more mainstream and the underlying technology becomes more transparent, the allure of connecting geospatially distant researchers, teachers, students, visitors, and community members through virtual environments is expected to attract even more learning organizations across sectors including corporate, public, nonprofit, and educational institutions.

References

Au, W. J. 2008. *The making of Second Life*. New York: Harper Collins.

Bainbridge, W. 2007. The scientific research potential of virtual worlds. *Science*, 317(July): 472-476.

Bates, B. 1988. *Information as an economic good: Sources of individual and social value*. University of Madison Wisconsin Press.

Bloomfield, R.J. 2007. *Worlds for study invitation: Virtual worlds for studying real-world business (and law, and politics, and sociology, and...)*. Retrieved July 4, 2009, from SSRN: http://papers.ssrn.com/sol3/papers.cfm?abstract_id=988984

Castells, M. 1996. *The rise of the network Society*. Malden, MA: Blackwell Publishing.

Castronova, E. 2005. *Synthetic worlds: The business and culture of online games*. Chicago: University of Chicago Press.

Choo, C.W. 1998. *The knowing organization: How organizations use information to construct meaning, create knowledge, and make decisions*. New York: Oxford University Press.

Deeley, L. 2007,March 24. Is this a real life, is this just fantasy? *Times Online UK*: Times Newspapers Ltd. Retrieved July 5, 2009, http://women.timesonline.co.uk/tol/life_and_style/women/body_and_soul/article1557980.ece

de Nood, D., and J. Attema. 2006. *Second Life: the second life of virtual reality*. The Hague: Electronic Highway Platform.

Drucker, P.F. 1989. *The new realities: In government and politics / in economics and business / in society and world view*. New York: Harper and Row Publishers.

Fetscherin, M., and C. Lattemann. 2007. *User acceptance of virtual worlds: An explorative study about Second Life*. Rollins College: University of Potsdam.

Franklin, U. 1999. *The real world of technology*. Toronto, Canada: Anansi.

Gee, J.P. 2007. Are video games good for learning? In *Worlds in Play*, ed. S. de Castell and J. Jenson, 323-336). New York, NY: Lang.

Jarmon, L. 2009, in press. Homo virtualis: Virtual worlds, learning, and an ecology of embodied interaction. *International Journal of Virtual & Personal Learning Environments*, 1(1).

Jarmon, L., and J. Sanchez. 2009. The Educators Coop: A model for collaboration and LSI communication research in the virtual world. *Electronic Journal of Communication*, 9(1&2). CIOS. Retrieved July 4, 2009, from http://www.cios.org/www/ejc/v19n12toc.htm

Jarmon, L., and J. Sanchez. 2008a. The Educators Coop: A virtual world model for real world collaboration. In *Proceedings of the Annual Meeting of the American Society for Information Science & Technology (ASIS&T)*. Columbus, OH (October 24 – 29, 2008). Retrieved July 4, 2009, from http://www.asis.org/Conferences/AM08/posters/82.html

Jarmon, L. and J. Sanchez. 2008b. The Educators Coop experience in Second Life: A model for collaboration. *The Journal of the Research Center for Educational Technology* 4(2): 66-82.

Jarmon, L., T. Traphagan, J. Traphagan, and L. Jones-Eaton. 2009, in press. Aging, lifelong learning, and the virtual world of Second Life. In *Higher Education in Second Life*, ed. C. Wankel and J. Kingsley (Charlotte, NC: Information Age Publishing, Emerald.

Kolb, A.Y., and D.A. Kolb. 2009. The learning way: Meta-cognitive aspects of experiential learning. *Simulation & Gaming*, 40(3): 297-327. Retrieved July 4, 2009, from http://sag.sagepub.com/cgi/content/short/40/3/297

Lave, J., and E. Wenger. 1991. *Situated learning: Legitimate peripheral participation*. New York: Cambridge University Press.

Linden, J. 2008. *Jack Linden Office Hours in Second Life*. 10/30/2008, 11:30 SLT. Retrieved July 4, 2009, from http://educatorscoop.org/sanchez/LTR/openspace.html

Linden Lab. 2009. *How meeting in Second Life transformed IBM's technology elite into virtual world believers*. Case Study. Retrieved on June 28, 2009, from: http://secondlifegrid.net.s3.amazonaws.com/docs/Second_Life_Case_IBM.pdf

Linden Lab. 2008. *Company Fact Sheet*. Retrieved July 5, 2009, from http://lindenlab.com/pressroom/general/factsheets/lindenoverview

Lynch, M.A., and R.J. Tunstall. 2008. When worlds collide: Developing game-design partnerships in universities. *Simulation Gaming*, 39(3): 379-398.

Mangan, K. 2008, January 11. Virtual worlds turn therapeutic for autistic disorders. *Chronicle of Higher Education*, 54(18):A26. Retrieved July 5, 2009, from http://chronicle.com/weekly/v54/i18/18a02601.htm

Mayrath, M., J. Sanchez, T. Traphagan, J. Heikes, and A.

Trivedi. 2007, June. *Using Second Life in an English course: Designing class activities to address learning objectives.* Paper presented at ED-MEDIA World Conference on Educational Multimedia, Hypermedia & Telecommunications, Vancouver, Canada.

Nardi, B. 2005. Beyond bandwidth: Dimensions of connection in interpersonal communication. *Computer Supported Cooperative Work*, 14(2): 91–130.

Reigeluth, C. 1999. *Instructional-design theories and models: A new paradigm of instructional theory* (Vol. 2). Mahwah, New Jersey: Lawrence Erlbaum Associates.

Sanchez, J. 2007. A sociotechnical systems analysis of Second Life in an undergraduate English course. In *Proceedings of World Conference on Educational Multimedia, Hypermedia and Telecommunications* 2007, ed. C. Montgomerie and J. Seale, 4254-4258. Chesapeake, VA: AACE. Retrieved July 4, 2009, from http://www.editlib.org/?fuseaction=Reader.ViewFullText&paper_id=25990

Stein, R. 2007, October 6. Real hope in a virtual world. *Washingtonpost.com*, p. A01. Retrieved July 5, 2009, from http://www.washingtonpost.com/wp-dyn/content/article/2007/10/05/AR2007100502391.html

Steinkuehler, C., and D. Williams. 2006. Where everybody knows your (screen) name: Online games as 'Third Places.' *Journal of Computer-Mediated Communication*, 11(4): 885-909.

Suchman, L. 2002. Practice-based design of information systems: Notes from the hyperdeveloped world. *The Information Society*, 18: 139–144.

Suchman, L. 1987. *Plans and situated actions: The problem of human-machine communication.* New York: Cambridge University Press.

Taylor, T.L. 2006. *Play between worlds: Exploring online game culture.* Cambridge, Massachusetts: The MIT Press.

Uhlenbruck, K., M. Kitt, and K. Meyer. 2003. Organizational transformation in transition economies: Resource-based and organizational learning perspectives. *Journal of Management Studies*, 40(2): 257-275.

Wenger, E. 1998. *Communities of practice: Learning, meaning, and identity.* New York: Cambridge UP.

Wenger, E., R. McDermott, and W.M. Snyder. 2002. *Cultivating communities of practice: A guide to managing knowledge.* Boston: Harvard Business School Press.

Williams, D., N. Ducheneaut, L. Xiong, N. Yee, and E. Nickell. 2006. From tree house to barracks: The social life of guilds in World of Warcraft. *Games and Culture*, 1(4): 338-361.

Yee, N. 2007. Motivations of play in online games. *Journal of CyberPsychology*, 9: 772-775.

Zuboff, S. 1988. Managing the information organization. In *The information society reader*, ed. F. Webster, 312-326). London: Routledge.

Mobile Phones for Informal Science Center Learning: A Socio-Technical Analysis

JAMES E KATZ, KATE HALEY GOLDMAN
& SUSAN FOUTZ

Rutgers University

The mobile phone is arguably the greatest success story in the history of communication technology: in less than a generation, the technology has gone from zero to more than four billion users around the world. Understandably, given the mobile phone's enormous popularity and seemingly non-stop use by people of all ages and backgrounds, science centers and museums want to consider how to use the technology. The excitement in the museum community over the mobile phone's potential for free-choice science learning included one of America's leading centers, the Liberty Science Center (LSC), which is located in Jersey City, New Jersey, across the Hudson River from New York City. In this chapter, we review our perspective of the way museum/science center visitors generally, and LSC visitors in particular, have responded to mobile-device-driven opportunities to engage in content. Needless to say, the views expressed are our own and do not necessarily reflect those of any institution.

Background and data collection

In 2006, LSC received a grant from the National Science Foundation (ISE-#0610352) for the *Science Now, Science Everywhere* project. *Science Now, Science Everywhere* yielded the acronym *SNSE*, which was pronounced *sensei*, meaning teacher or mentor in Japanese. *SNSE* was designed to explore the unique learning opportunities that are possible when visitors use their mobile phones as tools for learning in a science center environment. Specifically, LSC believed that visitors, by using their phones, would have more intellectual commitment or sense of ownership over their learning experiences. This in turn was believed to lead to continued use of the content they had acquired from the science center after their LSC visit. *SNSE* was designed as a mobile learning initiative to enable the public to interact with exhibits, retrieve content, and extend their learning by using their mobile phones. The program aimed to encourage visitors to use *SNSE* both on-site at the center and off-site. LSC developed a variety of mobile activities, including:

- Dial-in audio content
- Texting to an exhibit interactive
- Texting to receive web links
- Saving of designs created at the museum, and a
- Camera phone challenge.

A fundamental tenet of the *SNSE* project was to make use of the visitors' own personal technology devices, specifically their mobile phones, as an extension and enhancement of the free-choice learning that would stem from the LSC's resources and exhibits. By asking the visitors to use their own devices, rather than lending them devices to use during their visits, LSC staff members hypothesized that they could increase visitor interaction with the content material after the visitors have left the institution. Through using their own devices, their sense of intellectual ownership over the content material might increase. In addition, if visitors were able to share content with others, it might encourage post-visit review and reflection on the content material. All of this would serve to enrich the visitors' depth of knowledge and appreciation of science, technology, and related fields.

LSC underwent a major renovation and expansion during the *SNSE* grant period, reopening in July 2007. During the 22 months the facility was closed, a prototype of *SNSE* was installed in the LSC's temporary facility at its nearby facility in Riverside. This prototype offered audio content to a predator/prey exhibition entitled *Eat and Be Eaten*. When the refurbished main facility reopened, *Eat and Be Eaten* was reinstalled at the main LSC site and included *SNSE* audio content. Two other exhibitions at the renovated building were also designed to include *SNSE* components: *Communication*, which focuses on the evolution of human language and the ability to communicate abstract ideas, and *Breakthroughs*, a small multimedia temporary exhibition space. In the months following the re-opening of the building, LSC incorporated other *SNSE* audio stops throughout the building. In total there were 36 audio stops, one text-in bookmarking feature, two text-in exhibit interactives, one multimedia messaging service (MMS) exhibit interactive, and the camera phone challenge.

LSC drew upon two groups to help advance understanding of the potential uses of mobile communication in their facility with an eye towards enhancing opportunities for informal science learning. The Rutgers University Center for Mobile Communication Studies, led by James E. Katz, was the Research Partner on the *SNSE* project, and its work focused on teen use of mobile phones in general, and of *SNSE* in particular, at LSC. Data was collected at several junctures. The Research Team monitored three community evening events during which time families whose children were from economically disadvantaged school systems could visit the museum free of charge. Observations were also made of students on field trips to the science center. More general observations were made of visitor experiences, including post-visit interviews and visitor movements on the exhibition floor. The team also visited other museums and science centers to form a baseline of mobile phone use in informal learning environments.

The Institute for Learning Innovation (ILI) team, led by Kate Haley Goldman, conducted front-end, formative and summative evaluation of the *SNSE* project. The formative phase of the evaluation was aimed at providing periodic input throughout the project's cycle of development to help with suggestions for refinement. The summative evaluation was designed to compare actual use of exhibition components, overall enjoyment, self-perceived learning, interest in science and technology and post-visit use of and connection to the material. The mixed-methods design was constructed to observe visitors actually using *SNSE* with the exhibit components, interview both *SNSE*-users and non-users as they left the museum, and interview them again six or more weeks after their visit. Thus the investigators would be able to compare users and non-users both on exit and then six or more weeks later. (The exit interview process is shown in Fig. 2; data collection by ILI observer is shown in Fig. 1.)

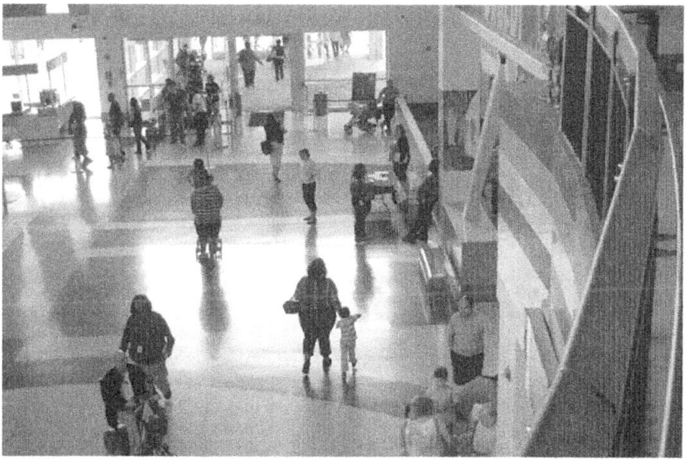

Figure 1: Exit interview process arrangement (Source, James E Katz)

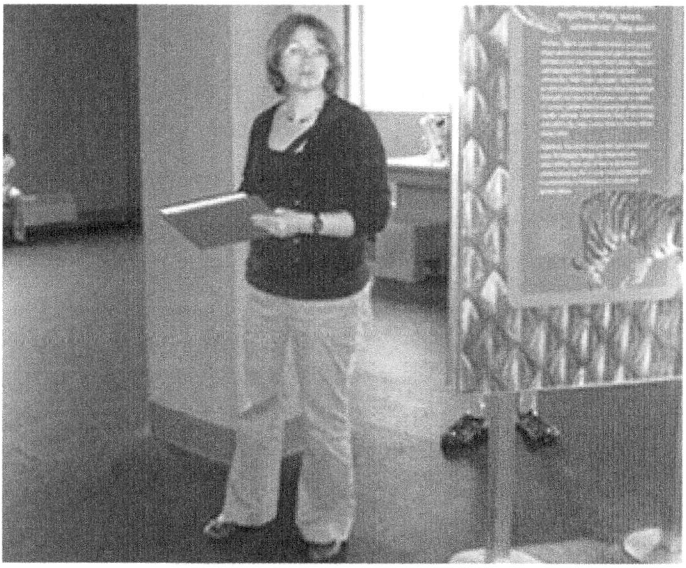

Figure 2: Observational data collection by ILI observer (Source: James E Katz)

Specific methods used in the summative evaluation included exit interviews at LSC, follow-up telephone interviews with those who had completed exit interviews, visitor observations at *SNSE* exhibits at LSC, and analysis of data collected by LSC through the back-end of the *SNSE* exhibits themselves.

Exit interviews were conducted at LSC from August 15–29, 2009 on both weekdays and weekend days, primarily in the afternoon. A total of 158 exit interviews was completed; 59% of the exit interviews were with non-*SNSE* users and 41% were with *SNSE* users. Due to the sampling method used, *SNSE* users were interviewed at a higher proportion than they naturally occurred in the science center.

Telephone interviews were conducted by ILI staff using the contact information provided by visitors on their exit interviews. Telephone interviews were conducted from November 20, 2009 through February 2, 2010. Two-thirds (66%) of those interviewed at the science center agreed to participate in the follow-up telephone interviews and provided contact information. Of those for whom contact information was available, 36% completed a telephone interview. In total, telephone interviews were completed with 38 individuals who also had participated in the exit interviews; 58% of the phone interviews were with non-*SNSE* users and 42% were with *SNSE* users. All visitors who completed a telephone interview received a $10 gift certificate for Amazon.com.

Visitor observations were designed with the goal of better understanding the experience of *SNSE* users and non-users at

SNSE enabled exhibits. Quantitative data included engagement level and time spent at the exhibit. Descriptions of the visitor's interaction with the exhibition and with other visitors or LSC staff were collected qualitatively. A total of 365 observations of visitors at *SNSE* exhibits were completed between August 15–29, 2009 on both weekdays and weekend days.

Science center visitors and the mobile phone

Are SNSE users different? Prior reviews of mobile phone use in museums have tended to focus on art and historical institutions, and much of the development of mobile phone functionality is currently directed for use within those institutions. (Center for History and New Media, 2010). To whatever extent there may be public expectations surrounding the use of mobile devices for free-choice learning in an institutionalized setting, it may well be that these are predicated on the art/historical model. This presumed identification and sense of utility may carry over to color user perceptions in a science-related environment. Hence, in the first instance, the *SNSE* project differs from traditional mobile free-choice learning in that it is situated in a science center, and thus the audience composition may be significantly different than in art and historical institutions.

Another factor to bear in mind is that it appears that most if not all visitors to LSC are there in a socially connected capacity, or to put it differently, there were very few, if any, solo visitors to LSC during the period of study. For example, all those at the science center with whom evaluation interviews were conducted turned out to be part of a group (the mean group size was 4.3 individuals). The majority of visitors (72%) were in a group that included adults and children (ages 12 and under). The group nature of LSC visitorship was further reinforced by observations during the *community evenings* at the LSC. (These were designed to bring school-age children, along with their parents/guardians, who were from New Jersey's economically disadvantaged schools to LSC. So while the point is not entirely surprising here, it does reinforce the fundamentally social nature of the visit to the science center.) Visitors tended to have an extended stay at the science center: with 55% spening more than three hours and 31% visiting for two to three hours. The majority of visitors interviewed (60%) had been to LSC previously. So while these data paint a picture of the LSC visitor, and highlight the socially embedded nature of the visit, it may well be that they vary from visitors to arts, history, or other cultural institutions. Unfortunately, we do not have data that bear directly on this point. However, the data does reveal that only 7% of those who were asked had used their phone at another museum before. *SNSE* users were no more likely than non-users to have used their phone at another museum.

Evaluation revealed few differences between *SNSE* users and non-users. There were no differences between *SNSE* users and non-users in the interviews based upon demographics, group type, or time spent in the science center. There was no significant difference between *SNSE* users and non-users in the observations based on crowding levels in the exhibition areas, age of the user, education level, group type, or previous visits to Liberty Science Center.

There were significant differences in *SNSE* users compared to non-users in a random subsample of the observations, with regards to the following factors (based on a sample size of 83, including 21 *SNSE* users and 62 non-*SNSE* users):

- Gender (p=.000). Males were more likely to be observed using SNSE than were females.
- Website use (p=.032). Those observed using SNSE were more likely to report that they had used the Liberty Science Center website than non-users.

There were several indicators that visitors who did use *SNSE* functions while at the LSC did so in a social way. Nearly half the visitors used some method to listen to the audio content in a shared fashion. This included, for instance, handing the mobile phone back and forth between a small group or family of visitors. For activities that involved texting information to the exhibit or sending created content, *SNSE* users were more likely to engage in conversation than non-*SNSE* users. Of those who used *SNSE*, the vast majority did not make use of the science content or share it with family and friends once leaving the institution. (At the same time, through interviews of visitors we did find numerous instances of young people using their mobile phones to take pictures of interesting exhibit items and then saving them for future sharing with friends or personal memories.)

Generational differences manifest themselves in people's

attitudes about phone use in public places. Many interviewees for the research study, especially parents, said that they felt they should not use their mobile phones in LSC unless there was an emergency situation. However, it is also important to understand that for some of these informants, using mobile phones to coordinate with other visitors in LSC would qualify as one of those emergency situations.

Researchers also observed that upon entry to LSC, and while still in the atrium, there were instances when school children were asked by their teacher to check whether they brought their mobile phones. If so, they were asked to turn them in for safekeeping during their visit to LSC. This seems to indicate that teachers have maintained normal school-based constraints on mobile phone use during their visit to LSC. At the same time, during the nominal collection process, some students were still using their phones to take pictures of the entrance area. This reinforces our belief that mobile phones do have significant promise as on "on-ramp" for informal science learning, albeit from a visual learning dimension. Perhaps there are even greater possibilities for the mobile phone in terms of individualized or personalized learning growing from the LSC experience.

Interestingly, this observation also raises a salient question as to whether there exists a gap between formal and informal learning contexts. Understandably, unlike the digitally minded younger generation, teachers and parents follow the conventional social norms to which they have long been acculturated, namely that there should be no phone use

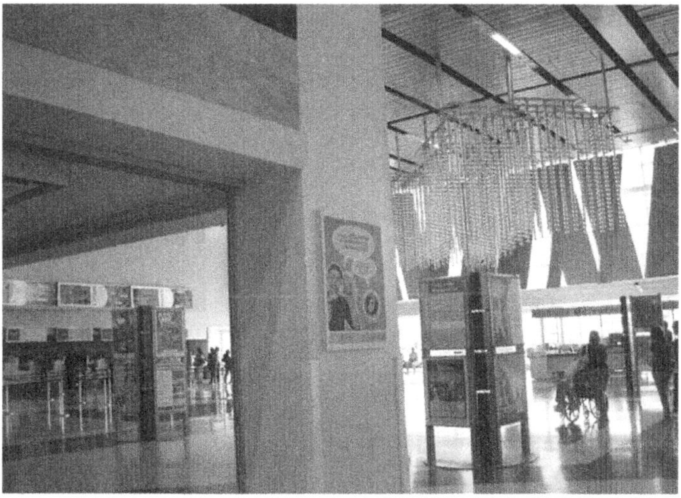

Figure 3: Display signs for SNSE in entry lobby (Source: James E Katz)

in public places, and especially not in museums. This appeared to be the reason that many visitors misinterpreted the signage. Although the signs were an invitation to use a mobile phone, a cursory glance by most visitors led them instead to interpret the signs as being a message proscribing use. (See Fig. 3, Display sign for *SNSE* in entry lobby, and Fig. 4, *SNSE* display sign in front of the *Eat and Be Eaten* exhibit.) This was because the message of non-use was already primed in the visitors' minds. Visitors perceived that mobile phone use in public places such as LSC was inappropriate because it might interrupt other visitors' experiences, or that they were barred from using them by museum policy. Here, researchers observed that social norms and social priming played a role among visitors' use, albeit the opposite of the way desired.

Does mobile phone use change the exhibit experience? There is some

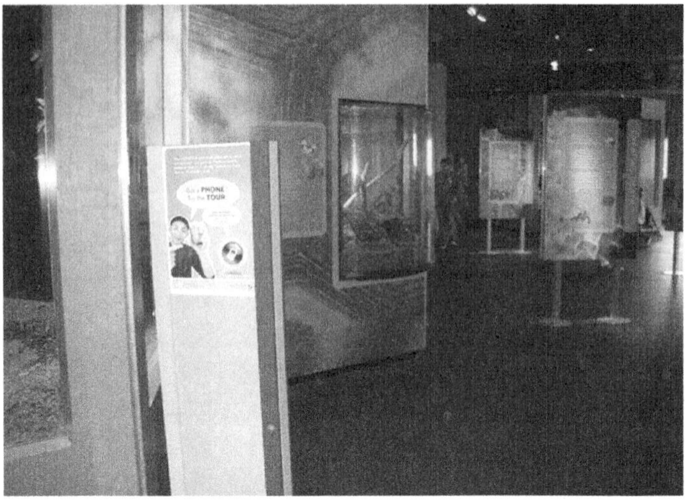

Figure 4: Display sign in front of the *Eat and Be Eaten* exhibit (Source: James E Katz)

evidence that visitors who used *SNSE* have a deeper interactive experience in terms of time spent. For the activities that involved texting the exhibit or receiving a picture from the exhibit, *SNSE* users were significantly more likely to complete the activity than non-*SNSE* users. Visitors who texted the exhibition component or used their phone to save content they had created showed increased interaction, including more pointing at the component and more social interaction. During the exit interview, *SNSE* users were more likely to report having learned something new while at LSC. They were also more likely to report having become more interested in science and technology during their visits, and having talked with others about science or technology during their visit.

Those changes however, did not necessarily mean that the *SNSE* users that we interviewed in the months after their visit generally remembered the *SNSE* activities. Approximately half of those interviewed had negative memories of the *SNSE* experience, either due to the audio quality or due to the confusing or dysfunctional nature of the activity. We do not have any evidence that the visit experience was more enjoyable for those who made use of *SNSE* functions. *SNSE* users did not report enjoying their visit more than non-*SNSE* visitors in the post-visit interviews. In recalling their use of *SNSE*, none of the users reported it was most memorable aspect of their visit.

Potential obstacles to visitor use: As mentioned above, there are still significant logistical barriers in implementing mobile phone use within a museum context. While the majority of those who attempted to access audio *SNSE* stops were able reach the content, hearing the content was a problem for many people. A third of those using an audio stop reported that they had difficulties hearing the content.

For those coming to LSC in the company of friends or family, there is sometimes a complex dynamism between interaction with distant others and the accompanying ("co-present") friends/family members. Apparently, there is considerable variation in how individual visitors choose to interact with co-present others *vis-à-vis* distant ones whom they are in contact with via the mobile phone. This has implications for potential interest in on-site mobile communication-based learning services.

There is also a dynamic relationship between involvement with the exhibits and with co-present others, due to childcare responsibilities, visitors feel that they cannot become involved in the often solitary investment of the *SNSE* service. One example, of this was when we encouraged a parent to use the *SNSE* activity, he kept reminding us, "Please help watch my kid." This example also highlights the dimension of interpersonal management which impinges on use of the mobile content.

Similarly, for some parents, a visit to an informal learning institution like LSC should be an occasion for parent-children interaction. In this situation, *SNSE* would seem to become a medium that works precisely against those with such a motive for visiting. Indeed, we observed a parent patiently teaching his kids about each exhibit in the *Eat and Be Eaten* area and also using a digital camera to take pictures of the animals. Hence, given these considerations, it may be that in terms of future technical development, that a learning activity would be more effective if it allowed for dyadic or group-level involvement (e.g., a visual guide downloadable on the mobile phone and a group of students can watch it on one student's phone). This echoes the finding noted above that nearly half of the visitors used the mobile content in a social way.

Yet the potential conflicts of different learning spaces are highlighted here not only in a conceptual sense. We found that physical settings are also an important factor that may affect mobile learning in LSC. Specifically, and as alluded to above, when people are in the virtual learning space using their mobile phones, the physical learning space is usually teeming with other visitors. We asked a few visitors for their evaluation of their initial *SNSE* trial, and received many responses that the ambient noises made it difficult to hear the exhibit descriptions through their mobile phones. Hence in terms of design considerations, it is important to incorporate the concept that visitors using *SNSE* might encounter a situation with conflicting learning spaces at several points.

Visitor social use: One important finding, possibly even due to the difficulties in hearing the audio content, was the social use of the material. Many audio stop *SNSE* users did put their phone on speaker to share the content with the entire group; 38% of audio stop users (14 out of 38) reported using their speaker phone function. Another 8% of audio stop users (3 out of 38) reported sharing the phone from person to person so that multiple people could listen to the content even though the phone was not on speaker. As a result of these two techniques, 45% of *SNSE* audio stop users listened to the audio content as a group.

Why do visitors seldom use their phones to interact with the exhibition?

One critical factor that had an impact on each of the potential outcomes was the extremely low use of *SNSE* overall, estimated in the summative report to be approximately 2% of the LSC visitation.

Non-users of *SNSE* were asked why they did not use their mobile phone to interact with the exhibits. Nearly a third of non-users (31%) were simply not aware that *SNSE* use was an option during their visit and another 9% did not have a phone with them during their visit. However, 60% of non-users were aware of *SNSE* but chose not to use it. Visitors who were aware of *SNSE* but did not use the system had diverse reasons for choosing not to do so, including:

- Using their phone in the museum was not appealing (44%)
- Being focused on child-care (17%)
- Not wanting to pay a fee for the call (15%)
- Not wanting to do a particular activity (9%)
- Being unclear about what the activity was (9%)
- Not knowing how to use their phone to do the activity (6%).

Despite the fact that provocative questions were asked as part of the signage to encourage mobile phone use as part of the *SNSE* exhibit, it was not clear to the visitor that the mobile phone-provided information was actually germane to what the visitor wanted to know about the exhibit. The stimulus to prompt the user to engage the service may not have been sufficiently compelling in light of the obstacles to use and the lack of social cuing to obtaining exhibit information via the mobile phone. Also, as noted above, the user is not primed either to interact with exhibits using the mobile phone or to gain information in this manner.

Low use is a barrier to understanding mobile phone potential

We still lack a full picture regarding what prompts visitors to use their mobile phone to interact with an exhibit, due to the low level of use. Awareness of *SNSE* functionality is an issue (nearly a third of the non-users were unaware of the *SNSE* capability) that better marketing and signage might address. Visitors had other significant concerns or obstacles that prevented them from participating. While for some visitors not knowing how to use their phone or not understanding the activity was a barrier, the majority of visitors understood the concept but were not interested in participating.

After reviewing *SNSE* and related mobile phone projects, we would observe that the low use rate is the most significant obstacle to both the collection of data and to deeper understanding of the experience of mobile phones in informal learning settings. An argument could be made that museums have failed to convince visitors that using their phone to engage with the institution will be worth the trouble. While there are not many published reports of the use rates of audio content in museums, the few accounts that exist suggest that in many institutions a 10% uptake rate is considered successful for audio content in a permanent collection (Proctor, 2010), although some institutions experience use rates as high at 17% (Luke and Stein, 2006). This is a primary issue, and understanding barriers to use for visitors remains one of the key issues for continued exploration.

Teens as early adopters for phones in the museum setting: the weakness of traditional descriptions of adoption patterns

In the grant, LSC hypothesized that teenagers were by nature, early adopters. To quote the original Co-Principal Investigator,

Denise Bressler, in her 2006 paper for the *Museums and the Web* Conference, "We believe that if we accept the teenagers' mobile culture and design learning activities specifically for them to use with their phones, young people will find the *SNSE* activities familiar and enjoyable." (Bressler, 2006)

The classic text in the study of the adoption of new ideas and technologies is Everett Rogers' book *Diffusion of Innovation* (2003), originally published in 1962 and most recently updated in 2003 to include phenomena such as video games. Rogers formulated distinct population segments including the Innovators, the Early Adopters, the Early Majority, the Late Majority, and the Laggards. Each segment has different characteristics and reacts to innovations in a different fashion. According to Rogers, innovators make up about 2.5% of the population, early adopters another 13.5%, followed by equal segments of the early majority (34%), late majority (34%) and then finally the last 16% of the population, the laggards. These percentages form a bell curve. In his model, adoption of innovation generally spread throughout the society on a rightward-stretched S-shaped curve, with one plateau early on as the innovation was changing, followed by a rapid escalation in users (the *tipping* point) followed by another plateau as the laggards eventually adopt the innovation. Although often called theory, the work of Rogers may be more accurately characterized as a set of inter-correlated variables that co-mingle psychological openness to innovation, economic calculus broadly defined, and perceived attributes of a given innovation.

In Rogers' *Diffusion of Innovation*, he explains that innovations vary greatly in their perceived attributes. These perceived attributes help explain why some innovations, such as mobile phones, are adopted quickly, while others, such as seat belts in cars, require decades to reach widespread adoption. These perceived attributes of the innovation, in combination with a variety of other variables (for example, nature of the social system, and communication channels), determine the rate of adoption of the innovation. The adopter categories have different values and characteristics.

The Rogers model has not been borne out in terms of mobile phone use for informal science learning, as the *SNSE* research team documented. Researchers concluded that while mobile phones were pervasive at LSC, they were not frequently used to interact with the content. To quote: "Specifically, we have the paradoxical finding that many people do use mobile phones extensively in the LSC, but not for the purpose of interacting with the exhibits or for informal STEM learning" (Katz, 2008). It might be said that the time span for adoption in museums and science centers has been short, and therefore society is still at the very early (and bottom) portion of what might be an extremely attenuated S-curve. That said, it is also the case that the mere presence of technology does not lead to its adoption, no matter how attenuated its S-curve might be. The problems we have seen in terms of the mobile phone technology nested within the science center have a great deal to do with other factors, including the perceived added value of the audio material, ambient noise conditions, and the social setting.

Other potential frameworks for understanding visitor use: One project assumption had been that possession of a mobile phone, and avid use of that phone as an early adopter, would translate into use of the phone to undertake particular activities within the science center space. Knowing who was an early adopter might have allowed museums to predict who would use this functionality in the future. Therefore the Diffusion of Innovation model could have been useful to other informal science education institutions if we could have operationalized early adoption within this context, helping the science education institutions understand the technology profile of their visitors. As the project progressed, we acknowledged there remains a substantial gap between adopting a technology (a mobile phone), particular features of that technology (texting) and the desire to use a given feature in every context. Texting is a means to satisfy a particular need, such as coordination of family schedules, reinforcing a sense of closeness with one's friends, or even killing time. However, despite the heavy use of texting in such contexts, it does not necessarily follow that users text in order to further their learning goals. One analogy would be to say that simply because one owns a car and likes to drive to certain places, one is *ipso facto* interested in driving to a given place.

During the formative evaluations of *SNSE* and through the Rutgers team research, ILI determined that the Diffusion of Innovation theory was not the best fitting model for looking at visitor adoption of these technologies in the museum. While we could at times characterize visitors into Rogers' categories, these seemed less predictive of actual use within the museum. As the evaluation developed, suggesting the original LSC hypotheses might not be borne out, ILI considered a number of other theories on the adoption of information technology besides the diffusion research. One approach ILI considered was the Technology Acceptance Model (TAM) as first described by Venkatesh and Davis (2000. See also Mathieson, Peacock, & Chin 2005, Ma & Liu 2004). It is possibly the most widely applied model of the acceptance of informational technology and is generally believed to be more predictive and robust than other models. In the TAM, perceived usefulness and perceived ease of use, in combination with a variety of external variables, are seen as the key factors in influencing attitudes towards the adoption of a technology. Those attitudes then contribute to the intention to use the technology and the actual use of the technology. Visitors might use their phones within the museum setting if they perceived the activity to be useful and/or they perceived the activity to be easy to use. Yet this proved very difficult for visitors in the formative studies to gauge, as the usefulness or ease of use of prototype or hypothetical prototype was too abstract.

Towards the end of the project, as we began to examine the results, we turned towards a framework known as the *Domestication Approach* to the adoption of technology (Haddon, 2003). As Ling articulates in his discussion of the domestication perspective, adoption is a process, centered in the everyday life of an individual within a particular context (2004). To adopt a technology, one needs to be able to imagine the use of that technology, appropriate it for one's own, and incorporate it into one's habits or practices. Domestication is not solely an individual process, but takes place often as a social interaction between individuals. TAM theories would suggest that visitors need to be able to imagine using the technology, understand why the technology would be useful to them, perceive it to be easy to use or worth using if complicated, and adopt the use of that technology into their practice.

Yet another perspective that could be taken on our results is that of *Apparatgeist*. This perspective, advanced by Katz and Aakhus (2002), emphasizes the cultural setting of the technology and the way in which status accrues to the user (or nonuser) of any given technology. Most important is the role of the technology in social relationships. Apparatgeist is more compatible with the perspectives taken with what is known as the social construction of technology, and is rather far from the Rogers model, which emphasizes, rather tautologically, the individual psychological characteristics of openness to innovation as predictors of innovativeness. Because the mobile phone is generally perceived as a way to communicate rather than a method to access information (at least until the advent of the iPhone and droid circa 2009), it is not entirely surprising that it would continue to be used this way by museum visitors. In terms of utilization patterns of young people, they tend to be extremely skeptical of institutional manipulation, but also extremely sensitive to social hierarchy and competition. Thus as an alternative or supplement to the "demand-response" method of information delivery, which characterizes *SNSE*, socially meaningful activities might be more usefully paired with the mobile phone. For example, socially engaging competitive games, especially those that involve physical exploration of the science center's space, might prove to be a fruitful direction for future development. Despite the absence of enthusiastic uptake at LSC, there seems to be little question that the mobile phone can be useful for informal science learning at several levels, ranging from coordination of group activities and sharing of assignment information to data collection and intellectual engagement. Perhaps the key is to find the methodology and content that works with the social interactive interests of the center visitors, rather than adding additional narrative layers to the information which is offered at the exhibits themselves.

Social connectivity and environment

An over-arching conclusion of our analysis to date is that mobile phones are heavily used by visitors to LSC, but it is much more for spontaneously connecting with the visitor's social networks and for coordination purposes than for informal learning as defined by the *SNSE* project. However, the seemingly paradoxical finding of heavy use for social networking and coordination, but much less for exhibit interaction, makes sense from a social interaction perspective. It is rather the case that most visitors are immersed in a rich social network of relationships with other people. Moreover, given the evolution of communication technologies, including the proliferation of mobile phones, visits to the museum or science center are increasingly taking place against a social background that requires coordination and contact. This is due to the fact that if a technology allows communication, it becomes incumbent on members of the social network to allow each other to be in contact with one another, and this is most especially the case of families that have childcare responsibilities. Upon entering LSC, the visitor is not cleaved off from these external forces; communication technology now allows or even seems to demand a form of perpetual contact with socially relevant but physically distant others.

In fact, this high level of mobile phone connectivity for visitors to museums can disturb other co-present visitors. It seems that this is a motive behind no-mobile-phone policies of some museums. We noticed in other museums that even when no-mobile-phone policies are prominently posted, users (and sometimes guards) furtively use their phones. Moreover, use is seemingly becoming normalized among visitors despite the attempts by museum authorities to control such use.

This may be illustrated by drawing on our comparative analysis of some New Jersey and New York museums. Observations were conducted at area museums in the fall of 2008 to provide a rough comparison point for understanding museum visitor behavior *vis-à-vis* mobile devices. During this observation period, several verbal duels were witnessed between guards and patrons over mobile phone use. Despite prominent supporting signage that was visibly posted in many exhibit areas, guards had to frequently prompt visitors who were using their phones in forbidden areas to stop using the devices. Sometimes users responded by challenging the guards, who had to provide arguments to the visitor as to why mobile phone use was not allowed. In one case, a shouting match ensued and the guard had to threaten to issue a citation to a recalcitrant mobile phone user. These incidents suggest that the fulcrum of normative behavior is moving towards museums having to justify their strictures on phone use to their visitors, rather than the other way around. This point is significant for the following reason: visitors are seemingly increasingly accustomed to using their mobile phones for coordination and communication with socially significant others. By contrast, even though visitors are well accustomed to these social uses, having practiced them extensively in the ordinary routines of their lives prior to visiting LSC, they have little or no experience or expectation concerning using phones to interact with ambient information resources. Thus there is no behavioral priming or baseline that would encourage users to begin using their mobile phones for informal science learning, even should there be initial interest in using their mobile phones to do so.

Despite the non-obvious connection of the mobile phone to informal science learning, there is clearly enormous enthusiasm for mobile phone use among visitors to LSC. We observed that phone use (voice as well as message checking) is quite common in LSC, especially near the entrance area and as visitors move from one exhibit area to another. Also, visitors take photographs of exhibits using their mobile phones (and of course with digital cameras as well), though the frequency varies considerably by exhibit.

Going beyond the content of exhibits themselves, we would like to highlight at a conceptual level the contrasting use patterns for mobile phones: frequent use for external communication versus infrequent seeking of additional learning opportunities via exhibits. This contrast provides an opportunity to consider how the *SNSE*-related technology could be modified to the findings to date; we pointed to some evidence which suggested that individuals' mobile learning in LSC may be related to the nature of and relationships to accompanying visitors. Attention could therefore be profitably directed towards helping visitors understand the new paradigm of mobile phone interaction with the exhibits that is being fostered at the LSC, but done in such a way that is responsive to the socially embedded situation in which visitors find themselves. This of course also goes to the heart of differential social standards regarding appropriate mobile phone use. Lastly, attention should be directed to the problem of interference that visitors may experience. These sources of interference may come both from areas surrounding the exhibit spaces as well as from others within the space. Much of the user feedback we received indicates that physical settings and interaction with co-present others are important dimensions that can work to reduce the use of mobile devices related to *SNSE*. Sources of interference are manifold, clearly, and many were outside the control of LSC. Yet, to the extent possible, consideration should be given to how to minimize various sources of interference on mobile learning.[1]

Conclusion and future directions

In summary, data shows that most visitors to LSC have their phones with them (if allowed by authorities). This fact serves as a good foundation upon which a *SNSE*-facilitated learning environment could be constructed. Yet during our studies we uncovered some factors that may account for varying uptake of *SNSE*. These factors have differential effects, but in particular, seem to be interfering with possible uptake of mobile phone-related services. Our analysis suggests that informal learning should be socially embedded in people's interaction behaviors and patterns of technology use, taking advantage of visitors' interest in sharing the content. It should take into consideration the symbolic significance and social interactive dimensions of a trip to a science center, capitalizing on the memory-making that individuals do through picture-taking.

As to future directions, multi-person involvement and game-like offerings would seem to hold promise. Such interaction should work to make the mobile phone interaction shared and inclusive, as opposed to the reverse. It would also be worthwhile exploring the possibility of developing conflict-narrative material, which would allow interaction and participation via mobiles. Having material that engages visitors with the physical space of the center via mobile phone contests or search activities could also prove to be a fruitful direction.

Greater consideration should be given to the physical limitations of mobile phone-based services within the science center environment. Though these limitations are often intractable, they nonetheless remain an important area for design improvements as future developments are considered. Sound cancelling technology, for instance, holds much promise in this area.

But perhaps the most important question is: Why bother? Unless there is a reason to use the mobile, most people will not. Compared to the wild excitement that accompanied the introduction of the Motorola flip phones two decades ago, and the more recent enthusiasm that greeted Apple's iPhone, the novelty of the mobile phone itself has been severely diminished. Hence people will need some motivation to go to the effort to access content, interact with, or otherwise become involved with science center offerings and ideas. These motives usually come from the ability to gain resources, be entertained, or compete socially. We have suggested a few ways the exhibits could be structured to tap into these motives. Yet, that said, it is also important to consider the rapidly changing nature of communication technology, so investments need to be moderated in order to be flexibly and quickly upgraded.

In conclusion, the mobile phone is not the hoped-for panacea for science centers and museums, but then neither was the video display, the computer, the Internet, or even television. Each of these has its place, and each remains in dynamic tension with the ever-changing demography and interests of museum visitors, as well as potential visitors. Museums and science centers are continually evolving towards greater sophistication in terms of both understanding of their visitors and the refinement of their exhibits. For their part, mobile technology and delivery platforms will also continue to advance and they, like other popular communication technologies before them, will surely find a place in the evolutionary trajectory of museums and science centers.

Notes

1. Beale (2006) delineates up to eight forms of mobile technology interference, including the tension with privacy/security issues, interference from physical device characteristics, enhanced feelings of social inclusion/exclusion within a group, changed perceptions of trust and affordances, interactions between formal and informal learning, environmental awareness and attentional control, conflicting old and new skills, and different expectations for technology use among different generations.

References

Beale, R. 2006. How to enhance the experience without interfering with it? In *Big issues in mobile learning: Report of a workshop by the Kaleidoscope Network of Excellence Mobile Learning Initiative*, ed. M. Sharples, (pp. 10-14). UK: University of Nottingham. Retrieved December 22, 2008, from http://telearn.noe-kaleidoscope.org/warehouse/Sharples-2006.pdf

Bressler, D. 2006. Mobile phones: A new way to engage teenagers in informal science learning. *In Museums and the Web 2006: Proceedings*, ed. J. Trant and D. Bearman. Toronto: Archives & Museum Informatics. Published online March 1, 2006 at http://www.archimuse.com/mw2006/papers/bressler/bressler.html

Center for History and New Media. 2010. *Assessment of the Field*. Retrieved from http://chnm.gmu.edu/labs/mobile-for-museums/assessment-of-the-field/.

Falk, J. H., and L. D. Dierking. 2000. *Learning from Museums*. Walnut Creek, CA: Altamira Press.

Haley Goldman, K. 2007. Cell Phones and Exhibitions 2.O: Moving beyond the Pilot Stage. In *Museums and the Web 2007: Proceedings*, ed. J. Trant and D. Bearman. Toronto: Archives & Museum Informatics, published March 1, 2007. http://www.archimuse.com/mw2007/papers/haleyGoldman/haleyGoldman.html

Haddon, L. 2003. Domestication and Mobile Telephony. In *Machines that Become Us: The Social Context of Personal Communication Technology*, ed. J. Katz, 43-56 . Piscataway,

NJ: Transaction Publishers.

Katz, J. 2008. *Informal science learning: SNSE summary report for 2008*. Internal Report submitted to Liberty Science Center.

Katz, J. and M. Aakhus, eds. 2002. Making Meaning of Mobiles. In *Perpetual Contact: Mobile Communication, Private Talk, Public Performance*, ed. J. Katz and M. Aakhus, 301-19. . Cambridge: Cambridge University Press.

Ling, R. 2004. *The Mobile Connection: The Cell Phone's Impact on Society*. San Francisco, CA: Morgan Kaufmann Publishers.

Luke, J. and J. Stein. 2006. *Walker Art Center: Summative Evaluation of Interpretive Experiences Newly Installed within the Permanent Collection*. Unpublished Technical Report. Institute for Learning Innovation. Annapolis, MD.

Ma, Q., and L. Liu. 2004. The Technology Acceptance Model: A meta-analysis of empirical findings. *Journal of Organizational and End User Computing*, 16(1): 59-72.

Mathieson, K., E. Peacock, and W. Chin. 2001. Extending the Technology Acceptance Model: The influence of perceived user resources. *The DATABASE for Advances in Information Systems*, 32(3): 86-112.

Moore, G. 2002. *Crossing the Chasm*. New York, NY: Collins.

Proctor, N., 2010. The Museum Is Mobile: Cross-platform Content Design for Audiences on the Go. In *Museums and the Web 2010: Proceedings*, ed. J. Trant and D. Bearman. Toronto: Archives & Museum Informatics. Published March 31, 2010. Retrieved from http://www.archimuse.com/mw2010/papers/proctor/proctor.html

Rogers, E. 2003. *Diffusion of Innovations*. New York, NY: Free Press.

Venkatesh, V., and F.D. Davis. 2000. A theoretical extension of the Technology Acceptance Model: Four longitudinal field studies. *Management Science*, 46(2): 186-204.

The Gyroscopic Museum

WAYNE LABAR

Liberty Science Center

Museums as we know them were born in a very different world than the one we inhabit today. Our world is filled with important new mediums and resulting social behaviors that expand and deepen the ways through which museums can achieve their missions. While challenging, these changes may also help a museum retain and improve its relevancy in visitor's lives. This chapter explores a new paradigm I describe as the Gyroscopic Museum.

Museums, while taking their name from the Greek word for the Muses, really began in the sixteenth century with cabinets of curiosity. These then moved to what are considered the first museums – open only to the upper and middle classes. It was not until the eighteenth century that the first public museums were established in Europe. These first museums were art-based, but their informal learning environments quickly gave rise to science, history, and cultural heritage institutions.

Looking at the technological and communication contexts in which these museums were founded, museums took advantage of state-of-the-art means to present information to the public. Primarily these were:

- Objects – authentic, real items of significance in all areas of human endeavor and the natural world related to the subject matter presented by the museum. Those of significance could be considered artifacts.
- Display – for the purposes of this discussion, a physical three-dimensional construct designed and fabricated to aid in the presentation of the content

and, at times, to protect the objects. At first these were mainly just display systems for objects.

- Graphic – textual information that presents information about the subject of the exhibition. Early on they were, for the most part, a means for identifying or organizing the objects.
- Live interpreter – the individuals who help interpret the objects for people visiting the museum. This could have been through organized tours or perhaps wandering interpretation as visitors guided themselves through the museum. One can imagine that the first live interpreters were the owners of cabinets of curiosities; they may have been the first "floor staff".

In looking at these "four pillars", it is clear that the major mediums by which the museum field continues to engage the public have not changed very much in the intervening hundreds of years leading up to this century. Let's examine the same list of categories using various items that might be assigned to them today. One should note that the object remains the object, but what may have changed is its authenticity (another paper altogether!):

- Objects – replicas, casts, artifacts, contemporary objects, historical objects, etc.
- Display – simulators, interactives, immersive interactives, IMAX® films, immersive environments, computer kiosks, dioramas, audio tours, podcasts, etc.

- Graphic – instructional, interpretive, introductory, environmental, banners, floor, LED, projected, way finding, magazines, newsletters, etc.
- Live interpreter – floor staff, education staff, docents, demonstrators, the Museum Theater, etc.

From the perspective of a museum's components, evolutionary development has occurred in the materials, people, and processes used to provide information to visitors.

The vast majority of these new methods are innately employed by staff in ways similar to those during the very early days of the museum field. Their method of use can be represented by the diagram at Fig.1.

Most of these new developments are being used in the same way as the original forms, in what can be called a radial model or a Radial Museum. By this I am referring to the following common characteristics:

- Unidirectional – The museum is pushing information out to the visitor; the museum is seen as the source of information, and the visitor is the recipient.
- Controlled – The information is created and generated by the museum, and as a result the information is controlled by the museum.
- Limited Resources – Empowering the first two aspects of this relationship is the fact that the museum has or controls almost all of the resources involved in the relationship between itself and its visitors.
- Location-specific – An inherent aspect of most

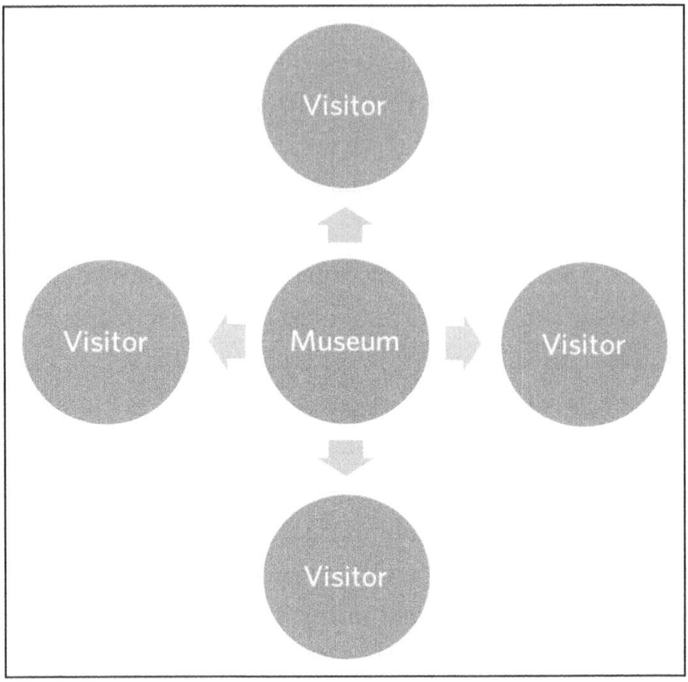

Figure 1: The Radial Museum

> of these mediums, coupled with the fact that the museum owns the mediums themselves, is that they are primarily tied to one physical location: the museum itself.

The result of the Radial Model is that while the museum does have control of the messages, intent, and focus, successfully achieving its desired impact is tied directly to the resources the museum can provide for delivering that message. In other words, the museum essentially controls the time, space, means, and message completely; therefore the success of the message lies with the museum. Faced with other organizations and forces with greater resources, it is amazing that museums have had such a large impact on our modern society. That being said, it is clear that the Radial Museum model is currently the modus operandi for many institutions.

This museum-centric model presents another significant hurdle to creating the societal impact that lies at the heart of the museum field. Remaining relevant – to a public that lives in a world where the inevitable march of time and development is changing society – becomes increasingly difficult, especially when the museum is primarily responsible for both the message and the means by which it is delivered. This is intensified by today's technological advances and their resulting consequences, which have quickened the rate of change we experience. With finite resources, museums face an uphill battle.

Ironically, the technological advances just described also offer an opportunity for museums to change the Radial Museum model into one that may make them increasingly relevant to their public and enable a broader impact. There are two innovations that together may bring about this change: the internet and the emergence of the mobile telephone into a mobile network node within the internet.

For the purpose of this discussion, let's first focus on the power of the internet relative to the museum model in two discrete areas: communication and personal creation. In terms of communication, the internet has afforded many people the opportunity to communicate back to the museum easily and for the museum to solicit input from the visitor. The internet facilitates this in terms of speed and opportunity. The directional flow of information suddenly changes, from one-way to two-way. Secondly, the internet has changed who supplies the information – with a force whose impact we are currently experiencing.

Lawrence Lessig, the father of the Creative Commons, in his book *Free Culture: The Nature and Future of Creativity* (2005) stated that, "Digital technologies, tied to the internet . . . could include a much wider and more diverse range of creators; those creators could produce and distribute a much more vibrant range of creativity." The world in which we live is increasingly fulfilling this potential. The internet has fueled perhaps the largest explosion of personal creation our society has ever seen, with the expanding world of blogs, Facebook pages, YouTube videos, opinion-offering opportunities, Flickr albums, peer-to-peer sharing, listservs, and other social media.

What this means for museums can be visualized in two dimensions by the diagram at Fig 2. This figure illustrates the concept that the internet facilitates communication and empowers personal creation in a world where museums not only send information to visitors, but also information is shared back with the museum. Additionally and perhaps even more importantly, this allows visitors to communicate and interpret experiences between themselves, without direct museum involvement.

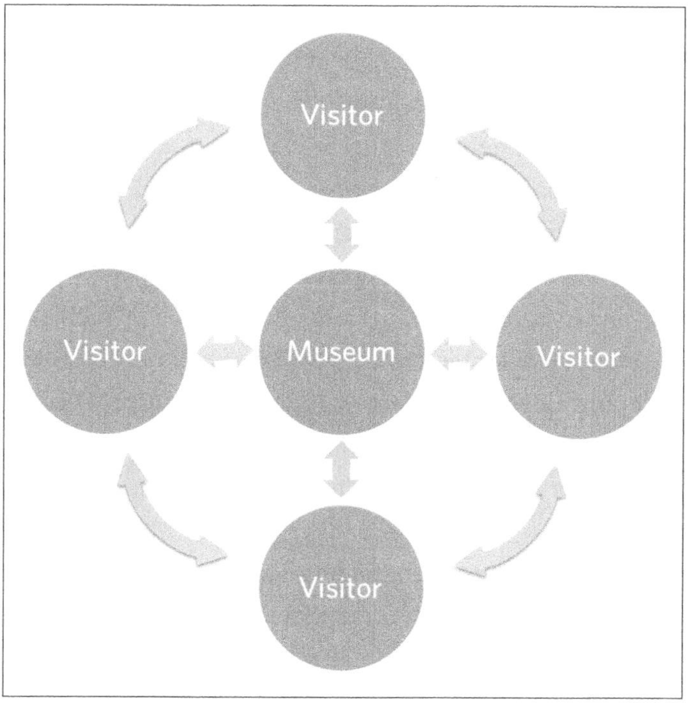

Figure 2: Internet Radial

The simplicity of this diagram belies the impact of the second technological aspect of society: the mobile device as a node on the internet. Cellular phones were first seen just as a means to share voice communications in the way physical land lines do. But just as people used modems to connect to the internet, it was a simple jump for people to think that cellular systems could connect them to the internet; a convergence of mobile technologies and the internet has rapidly taken place. New forms of mobile devices continue to emerge that expand the ways we can connect to the internet "mobilely", such as laptops, netbooks, eBooks, and tablets (the iPad is one well-known example of this technological species).

What is groundbreaking is that sharing content and interpretation between museum and visitor or visitor-to-visitor can happen at any time. Never before could a museum's mission impact people while on their way to work, in school, at the park, or in countless other places. In other words, museums have the opportunity to provide or facilitate mission-related interaction anywhere, anytime. The power of mobile devices increases with expanded capabilities – photography, music, and other innovative components – and it allows for a vast array of visitor-created content at any time as well.

Finally, we should not forget that in our model, visitors do not live in isolation between themselves and the museum. The internet is also a vast supplier of information that can be shared with an enormous audience. Informational veracity remains an elusive aspect, but clearly the internet has allowed large numbers of people access to an exponentially increasing amount of data. Home access – to the *New York Times*, the BBC, the Library of Congress, and museums' collections as they are digitized – has expanded the channels through which people receive information.

If we add what might be called the "three-dimensionality" of freedom of time and space offered by our mobile devices, as well as connections to the world outside of our visitor-only circle, our model significantly changes into what can be called the Gyroscopic Model (Fig.3).

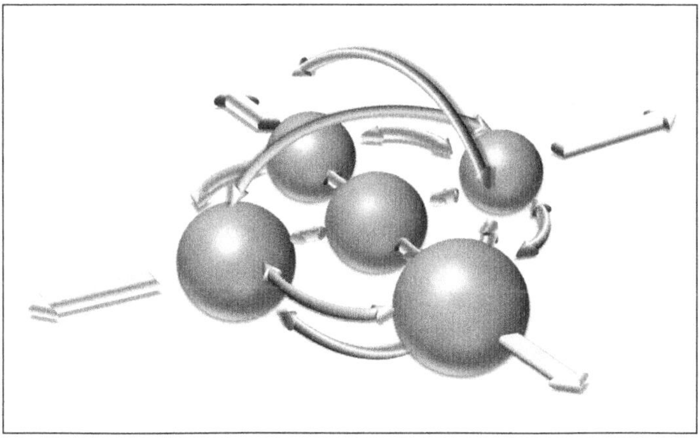

Figure 3: Gyroscopic Model

Here we have pictured not only the links between visitors and those relating to the museum, but also the connections to the outside world (the yellow arrows) at any time of day, including people who are not visitors to the museum. But why call it the Gyroscopic Model? Well, it does bear an uncanny resemblance to the gyroscope (Fig.4).

Looking closer, the gyroscope offers a singular model of what technological advances and the resulting change in relationships with our museum visitors (and even nonmuseum visitors) could allow a museum to do. This can be found in the definition of a gyroscope.

According to Dictionary.com Unabridged (Random House, Inc. 2006), a gyroscope is: "an apparatus consisting of a rotating wheel so mounted that its axis can turn freely in certain or all directions, and capable of maintaining the same absolute direction in space in spite of movements of the

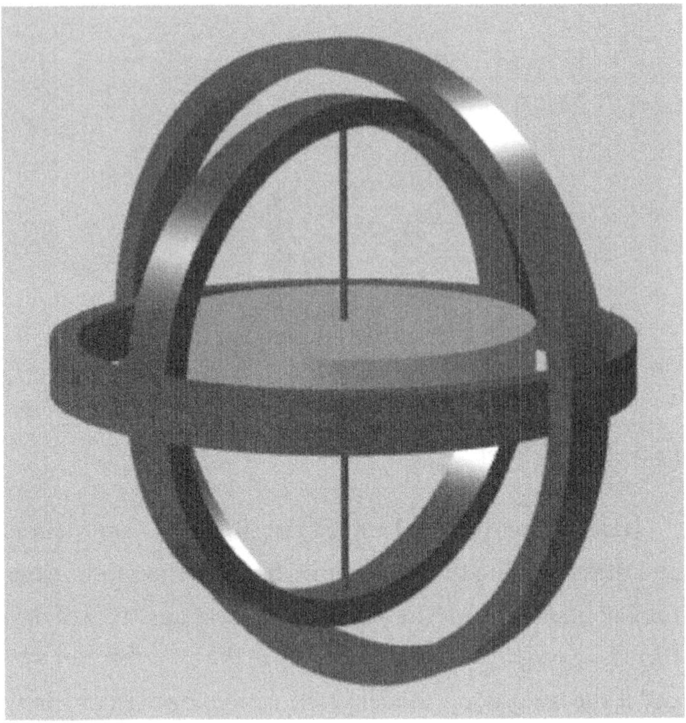

Figure 4: Gyroscope

mountings and surrounding parts."

Due to the physics involved, the result of the gyroscope's behavior is that once set in motion, by its very nature it tries to maintain a singular orientation. This has been critical in the use of gyroscopes to guide objects that experience forces from three dimensions – space craft, missiles, and airplanes, for example – and need to remain on course.

This behavior bears a striking resemblance to the goals to which museums aspire. Museums continually attempt to remain true to their missions despite societal forces that might affect them. Another element in this challenge is that museums now try to remain relevant to the public (from interests to social issues), as an ever-increasing number of changes impact society.

Based on how we arrived at the gyroscopic model as a representation of new technologies – the internet, social media and mobile devices that provide two-way connections between museums and visitors – the key parameters changing are the relationships. The desire for relevance and the empowering nature of relationship changes now occurring through the evolution of technology and software allow us to suggest a definition of what can be called the Gyroscopic Museum: "A Gyroscopic Museum is a museum consisting of relationships with its visitors that can respond freely in certain or all directions, and capable of maintaining being a relevant resource in spite of movements and changes in content and the surrounding world."

The key to remaining relevant is changing one's relationship with the visitor and the world in general, and in order to do so one must intelligently embrace the capabilities of this new technology.

Re-examining the common characteristics of the Radial Museum, we find that they change significantly under the Gyroscopic Museum model in terms of relationships. The Gyroscopic Museum is:

- Bidirectional – Now, instead of only the museum pushing information out to the visitor, visitors are supplying information and content to the museum, which then offers that content to other visitors. The museum is no longer seen as the sole source of information, and the museum is also now seen as a recipient.

- Uncontrolled and independent – Now that visitors are creating information, control of that data must be shared by the museum. Information from the world outside of the visitor-to-museum or visitor-to-visitor relationship plays a more important role. The result is that the relationships and information are not completely controlled by the museum and can actually be independent of the museum.
- Expanded resources – In the Gyroscopic Museum, resources that enable the operation of the organization are not completely owned by the museum. From the mobile devices that visitors own and use to the creativity and ideas they produce, a greatly expanded suite of resources becomes available.
- Location-independent – This last element completely changes location from the specificity of a physical museum to the wide reach of mobile devices, which allows for content and interpretation to be facilitated by the museum anywhere outside of its walls.

These are not easy changes for museums – they would be more comfortable completely controlling the message and understanding the roles of all contributors. Within the complexity of the modern world and the fact that many visitors expect to participate in companies and institutions, these technological developments will become a greater part of museum operations; the question is whether a museum will be able to use new technologies within the Gyroscopic Museum model.

Even if used as described, museums will still face a series of challenges arising from the outcomes, consequences, and changes that the Gyroscopic Museum model will produce. Assuming that museums do embrace this model, the way things operate could change significantly.

First, there is opportunity for a greater amount of interpretation, content, and mission-related achievements to occur, because the museum now has a wider range of available resources – including information, content creators, and physical assets. The challenge will be to utilize these assets effectively and also encourage the public to engage their property in this relationship.

A second dimension of using these resources is how to develop interactions with visitors, through physical or technology-driven encounters, that exponentially expand to nonmuseumgoers. Creating an experience with a visitor that results in a second or a series of experiences – which are then shared between that person and people who never visit the museum – is an incredibly powerful idea. It expands the reach of the museum and its relevancy to the public in general. Investigating ways to achieve this and determining the most effective approaches should be considered by the field as a strategic goal. Many museums measure success by attendance figures. The possibility that museums could have impact and achieve their mission while never seeing or personally touching an individual will question this widely accepted metric.

Thirdly, the mobile technology that allows the Gyroscopic Museum to exist means that the interpretation and information derived from the expanded resource pool mentioned above may be placed directly into context of the visitor's daily life. An example is the *Seafood Watch Pocket Guide* distributed by the Monterey Bay Aquarium, which helps people make sustainable seafood choices when shopping for food or dining at a restaurant. Museums should be exploring other ways to build on this to foster real world, decision-making moments.

Finally, the timeliness of the technology that empowers a Gyroscopic Museum and the created bidirectional relationships mean that topics relevant to the here-and-now are placed at a premium. This gives museums the opportunity to be proactive on issues and subjects directly related to our society: for science museums, it means playing a role in topics like global warming and energy; for history museums, current political events or clashes; and for art museums, looking at contemporary artists and cultural definitions. The question of course is whether museums have the courage to take on these issues. The answer is that they may not have a choice, for success in today's marketplace requires an institution to be relevant (as perceived by the general public) and to have the social significance often required by foundations and other funding sources.

In conclusion, the internet and its facilitation of social media and social media networks and the arrival of powerful mobile networking devices have fundamentally changed the relationships museums can have with visitors, nonvisitors, and the world at large. Museums have the opportunity to evolve from a radial model to a gyroscopic model by utilizing the very technological developments that brought about this change; they can then become more relevant and more adaptable to subsequent change. This will require redefining roles and accepting a different level of control. There are challenges and opportunities for museums that endeavor to embrace this new model, resulting in opportunities to place them at the core of society's most relevant concerns and issues.

For the past five years, Liberty Science Center has been involved in experimenting and exploring several aspects of the Gyroscopic Museum. They include:

- The Exhibit Commons – To determine whether we could use this same technology (and associated expectations) as a low-barrier process to have our visitors participate in the creation of exhibitions. Our investigation of engaging the public in the exhibition development process, called "Exhibit Commons." This started in 2005.
- Science Now Science Everywhere (SNSE) - The idea of engaging the public by having them use their mobile

devices within the museum setting

- *Cooking: the Exhibition* social media site – An experiment in opening up the complete exhibition process to the public through social media.

Beginning in 2008, Liberty Science Center became engaged in the exploration of how the latest aspect might work through the development and design of its next new experience, *Cooking: the Exhibition*. Because of wide cultural interest in cooking and food, the connections to all human beings, and a large professional infrastructure on the subject, the staff felt that this exhibition offered an opportunity to experiment with involving the public, since this potentially provided enough "meat on the bones" (Please pardon the food/cooking analogies, but they are hard to resist when talking about this project!). This effort fell into an ongoing series of investigations Liberty Science Center has undertaken – an attempt to engage the public in its exhibition process and operations, within an endeavor called the Exhibit Commons. While the experiment is in progress, this article lays out preliminary observations and discoveries made during the first year.

Observation One: Some subjects may lend themselves to public participation. On the other hand, all subjects probably lend themselves to "expert" participation, and with the creation of public "experts", perhaps any subject is game.

The structure behind this effort is an attempt to include the public in the process in which Liberty Science Center typically engages when creating an exhibition. Therefore, at this point we decided not to change the process (team meetings, research, etc.) but just to add the public as outside team members. Initially, we began to involve others in the project through a social media website. We used the Ning system since it offered some easy-to-use features that we could deploy immediately. The site is at www.cookingexhibtion.org and is known as *Cooking: the Exhibition Chefs* website. This was the launch point for all of the impacts to date and remains a key driver in changing and affecting what we do.

First, inviting anyone who wanted to participate on the team/in the project process meant that we were attempting to make the exhibition process transparent through using this social site – one of the principal ways to keep our "invisible" team members involved. This foundational underpinning of the site's importance lies at the heart of several significant observations.

A key advantage of the site is that is can be set up with no web programming or infrastructure requirements. It was literally created and online within one day. This flexibility has also allowed the team to respond quickly and on the fly to the needs that the process encounters. In our efforts for transparency and to include the public, we have worked hard to make this the **only** site where all comments, notes, research, and ideas have been posted. On the project management side, all drafts and working documents are posted and all meeting requests are made through the site. The problem we have encountered is that the social site is meant for socializing, not project management, and as a result, we are constantly looking for ways to improve its organization.

Observation Two: Using public social network programming offers the flexibility that is needed by the team, can be mastered by anyone with little effort, and makes it easy for everyone to participate. That also means one must live with the limitation that social media is not designed as management software. But more purposeful project management software may set too high a barrier for allowing the public to engage in the process.

When this process began, it was unclear who would be interested in participating in such an endeavor. While the idea of opening up the exhibition process is something the field has discussed and considered, "who will participate" is a fascinating question. Currently (as this goes to press) the *Exhibition Chefs* has 183 members and is growing at a steady rate of one to two members per week, with little to no advertising. Looking back at the "long tail" concept – small numbers of people being interested in every subject under the sun – team membership certainly reflects that idea. People join for different reasons, and opening it to everyone results in certain changes to the relationship between one's work and one's life. Categories of interest can be simplified into the following:

- Exhibition development and design (museum professionals)
- Renting the exhibition (museum professionals)
- Social media (museum and technology professionals)

- Cooking as pastime
- Cooking as profession
- LSC (members and visitors)
- Project members (family and friends)

While one might surmise that most members minimally engage in the project, a number of them participate actively: the excitement and energy that some of our members bring to the project, their willingness to participate, and their additional ideas and feedback have been invaluable to the exhibition. In fact, these have been so meaningful that some members have changed from being "the public" to becoming partners with potentially full-scale roles. On a related note, not only do public members provide feedback, but they have also taken on a more physical, real-world presence in unexpected ways. From actually showing up at the science center to participating in a meeting in person, to sending examples of food they have grown or cooked, this aspect has generated far more engagement for the full project team.

Observation Three: Not all of the public are the same. Project details are typically shared with those working on it and related experts, and inviting the public changes knowledge borders. Suddenly friends, family, and museum friends join social networks in unexpected ways. How does one respond if your mother criticizes your work in front of your professional colleagues and friends?

We all look to learn from one another in the museum field. Our profession has, in general, a long history of standing on the shoulders of those who have gone before. Meanwhile, ongoing subjects that will probably never disappear are questions around the ownership of ideas or design of exhibits and exhibitions. In addition, during exhibition development we often invoke confidentiality about ideas and proposals for competitive grant applications, designs that are in the conceptual stage, and other information. In order to engage all of our team members in the Cooking project, we have already posted early drafts of material concerning exhibit ideas, concepts, and a National Science Foundation (NSF) preliminary grant. Several members of the public reviewed the grant for this round and provided invaluable feedback.

Observation Four: One must carefully consider expectations and what the concept of transparency may mean. While others may become aware of certain aspects of the project, these need to be weighed against the combined assets this process can create.

The exhibition is now moving rapidly through conceptual design. This next phase will be one where, as in a typical exhibition project, members of the team may begin to concentrate on work in which they specialize and then move to present ideas together. While prior to this phase meetings occurred primarily at a weekly pace, the number of meetings between different team members will increase.

Working with the public will certainly need to be modified. We will need to answer questions of whether public team members can play any specialty roles, how to merge involvement in smaller meetings, how to share concepts and drawings, and what tools can be used in meetings to allow everyone to participate equally.

References

Lessig, Lawrence. 2005. *Free Culture: The Nature and Future of Creativity.* New York: Penguin Books.

Random House Unabridged Dictionary, 2006, s.v. "gyroscope." Dictionary.com Unabridged. Random House, Inc. http://dictionary.reference.com/

Conclusion

JAMES E KATZ, LORA APPEL
& ELLEN LYNCH

The chapters in this volume speak eloquently to the topic of social media and mobiles in the museum/science center setting. Yet it is worth reminding the reader that, cumulatively, they demonstrate the profound role that these technologies are increasingly playing in helping museums/science centers fulfil their increasing number of missions. These missions include preserving and fostering learning while seeking to engage new publics in the endeavor of learning itself. With increasing frequency, missions also include encouraging the co-creation of knowledge and insight even as they seek to contest received wisdom and traditional cultural meanings and interpretations. As they engage in these processes they are far from alone. Rather they are endemic of society-wide master trends. These include demands for accountability and efficiency, reciprocal inter-organizational constraints, an ever-expanding need to demonstrate social responsibility, and broader definitions of groups to be served. They must also grapple with both internal and external pressure not only to stay technologically current but also to devote resources to exploring and assessing innovations. Though many of the book's chapters deal with specific programs and initiatives in the museum/science center setting, one can look beyond them to these fundamental trends as propelling and guiding them.

As described by the authors in this collection, there have been some remarkable efforts in recent years by museums and science centers to engage their audiences and fulfil their mission using mobile media and social networking tools. We can discern three main strands. One has been to make their holdings and exhibits ever more available to audiences. Some segments of museum audiences have responded enthusiastically, but in many other segments much less so. The reasons for this are many, and will be addressed later. Another strand has been to get visitors and broad audiences engaged in responding to exhibit materials. In some cases, museums have sought not only interaction but also co-creation of materials by audience members. This in effect breaks down the walls between possessors of knowledge and recipients, a situation that can both enrich and problematize the museum experience and the role it plays in society.

In terms of social change, museums have sought to make their collections, and the insights of curatorial staff, ever more widely available to the public, and especially to those sectors that have traditionally either been overlooked by museums or themselves have ignored or by-passed such cultural institutions. At the same time, most museums feel beset with demands that seem unattainable and under assault from many quarters. These feelings of siege, most especially in economic terms, have come about even when attendance records are routinely being broken. In terms of technology, museums have grappled with new technology in an effort to connect with their visitors, both real and potential, and gain better command over their internal resources. While digital imaging and computer networking have done much in this regard, there is a gnawing worry that digital access and social media are critical to the future success, if not existence, of many museums. Yet, to the extent this view is correct, or at least perceived to be correct, serious questions arise as to the future of the bricks and mortar facility that physically embodies the museum. Disturbing questions exist not only about the physical nature of the museum but also its character. Doubtless for many museum enthusiasts, the digital world, and even more so the world of social media, are viewed as antithetical to the profound mission of the museum as a cultural institution. In some quarters, there has certainly been a temptation to perceive the museum as an institution that is not exclusively an adjunct to contemporary society, and therefore to be driven by transient whims – but rather part of a grand mission, an essential dimension of an intellectual tradition and program of cultural development that stretches from past centuries into the distant future. This view is strengthened by the too-frequent failure of social media to serve their intended purpose of firing up new (*new* as often defined as marginalized and young) constituents. Consequently, museums are beset with concerns and ambitions about social media, and particularly their adjunct of the mobile phone, since these add a new porosity to the museum experience.

The chapters included in this volume recognize the limited success, as well as the many difficulties, encountered when trying to harness the otherwise enormous potential of social technologies (i.e., social networking and mobile communication media) to engage museum audiences in a process of meaning-making. They also show how, when mixed with social technology, the existing assumptions and dispositions of audiences often leads to unintended or even surprising behaviors and understandings.

Not all the interest in social media and mobile springs from the interests of museum staff, though much does. There is substantial pressure from funders and regulators to use these technologies to meet priority items on the national agenda, most notably accessibility and inclusion. Some chapter authors call attention to the need to better understand the current generation of multi-tasking and active learners. This is necessary, *inter alia*, to design effective mobile and social media platforms. Authors Proctor and Murero suggest that while these emerging technologies might increase audience engagement, this active engagement inevitably must contain unpredictability at many levels, including behavior and content. In this regard, LaBar, Kahr-Højland, and Caron and Caronia show how it is possible to engage audiences through the active sharing of narrative both online and offline, and through the use of immersive physical interaction. Hence, better efforts could be made by institutions to analyze the large amount of digital trace data being acquired through mobile devices and social media, and for that matter, websites. The purpose of doing so should be to use these results to iteratively modify existing practices and seek to develop better content. That is, there is much that could be learned about the effectiveness of current efforts which, if systematically and comparatively analyzed, would generate potentially valuable insights. This could be especially valuable if conducted across institutional sectors.

As to more specific strands, we find the following among the chapter contributions.

Evolution of museums

Many of the authors begin their analyses by providing a conceptual overview of the museum terrain, analyzing how museums have changed over the years; they also put forward their thoughts concerning anticipated paradigm shifts that have taken place, and that they foresee in the future. Contributors Hug, Zimmerman, Sanchez, Ling and Kahr-Højland discuss the evolution from the traditional museum typical of the 1960s, which emphasized providing essentially static information to visitors. This stands in contrast to today's paradigms for visitors, which emphasize experiential and interactive environments. Free play is not only allowed but sometimes actively encouraged. Thus one might say that museums' orientation towards learning has shifted from a positivistic approach, which we can call Paradigm I, to Paradigm II, which we can say emphasizes a constructivist approach. Echoed in the chapters throughout this volume is the idea that museums are no longer spaces where education is disseminated unidirectionally but rather aim to be places that support open debate and critical reflection by means of interactivity and flexible narrative structures, all facilitated by digital technologies. The readily anticipated step, therefore, is Paradigm III, in which museums and science centers become institutions that still more strongly emphasize the interactive, exploratory, experiential, and even co-creative aspects of cognitive engagement with audience members. The engagement is in its most ambitious vision not only to be cognitive but psychologically involving and emotionally meaningful as well. Yet also part of this vision in Paradigm III is the notion of contestation and multiple layers of meaning, interpretation, and even reality. Lay involvement in Paradigm III goes far beyond the advisory board to the museum leadership or membership surveys, typical of Paradigm II. Rather, lay involvement is to penetrate to every level of the museum exhibition and curatorial undertakings. And the lay participation is no longer limited to those proximate to the museum, such as residents in the museum's locale. Instead, the invitation to openness and participation is global in scope and universal in orientation. This naturally fits the broader intellectual trend of extreme relativism and subjectivism that spans post-Modern circles. This trend challenges everything, and sees that all text or iconography, broadly conceived, is open to multiple contested interpretations, none of which can be considered authoritative. Museum and industry experts are in many cases energetically trying to create Paradigm III using social media and mobiles, among other products of the digital age. To create such an environment, museums must wrestle with the conundrum of trying both to introduce complex material and provide a degree of authority while also relinquishing enough control over content to empower outside contributors and co-creators.

The implementation of Paradigm III is neither simple nor uncontested. Endorsement of Paradigm III is by no means universal; even those who may well endorse Paradigm II sometimes recoil at Paradigm III. Nonetheless, both Paradigm II and III see themselves as increasingly dependent on social media, and increasingly the addition of mobility, to realize these ambitious goals. Yet even Paradigm I is alive and well. Often social media and mobiles are seen as a way to push material out to audiences, both new and old. Many within the museum community adhere to the belief that much knowledge is unchanging. Disciplinary expertise and boundaries are also important to many museum leaders. The essential notion of discipline, at its many levels of meaning, does not always fit easily with the freedom to explore and create. Hence, we can expect not only contestation among visitors about the museum exhibits, idealized under Paradigm III, but also among museum staff themselves about the right Paradigm to be followed. Nonetheless, such contestation with external as well as internal stakeholders remains for many museums a welcome challenge as they seek to find the right mix of activities to support the evolving vision of the museum.

Moving to a finer level of granularity, chapter authors also engage with issues of knowledge management in a complex and contested environment. Some friction naturally arises from well-meaning (and otherwise) attempts at cooperation among divergent parts of the system. (Here the system can be considered either a unit, such as a museum or science center itself, or a population of institutions within a geographic or conceptual boundary.) Several chapter authors acknowledge that many institutions would benefit from cost and knowledge savings as well as compiling a larger artifact library with less duplication. However, they would have to find new ways beyond current arrangements and agreements in order to better share their expertise. According to Sanchez, the experience with Second Life demonstrates that collaboration is one of the most significant and effective ways to use social media as a tool. But as Proctor points out, this kind of collaboration is generally hard to achieve, as personal differences in artistic and educational opinion held by institutions and curators, as well as the inherent notion of institutional rivalry, often prevail.

Fit between technology and mission of museums

Since the explosion of internet use and the more recent but high penetration of Web 2.0 applications and mobile technologies, institutions and organizations of informal learning, such as museums and science centers, have pushed to include points of interaction using such technologies within their respective markets. However, they have done so with remarkably little research into the fit between these emerging technologies and their mission. Despite several notable attempts, it remains to be seen how audiences embrace the ubiquitous nature of personal mobile technology (such as smart phones) in relation to the museum environment. In his chapter, Ling points out that those heavy iPhone users are not necessarily the ones who are likely to be engaging museums. Although studies show that iPhone users are the heaviest consumers of internet through mobile technology, research has also shown that these users rarely overlap with the current museum-goers, and of those that use museum-related mobile phone applications, few use these as originally intended. Hug in his analysis also describes this potential missed connection.

This highlights the fact that it is not necessarily the technology itself that is a barrier to increased public participation. Despite discussions of digital divides, access itself does not appear to be a major factor that prevents users from accessing museum-based information. Rather it is the cultural and social orientation as well as the implicit use of smart phone apps.

Katz, Haley Goldman and Foutz show that mobiles are heavily used to support the museum/science center visit, but much more as an enabling technology for coordination and relationship management rather than for content acquisition or investigation. Timpson points out that digital technologies are used currently merely as an outreach tool as opposed to an enhancement to the experience of visitors. However Zimmerman stresses that mobiles and Web 2.0 applications are in fact the key to life-long learning. They achieve this by facilitating something he calls "learning across contexts." Hug also supports this notion that gathering information from different sources in varying ways over a period of time can help students, as well as the general public, connect ideas between classrooms, literature and in-field learning. Moreover, Timpson and Sanchez, who agree with the idea of this untapped potential, argue that institutions ought to account for the ever-changing environment of digital technologies and social media trends, not only in exhibit design but particularly when building a museum or exhibit infrastructure. Museums should plan ahead to create flexible tools that can evolve and be re-used once current hardware becomes obsolete. Planners need to consider how the inherent structural design affects the user experience and how exhibits can be created with collaborative spaces intertwined within them.

On the other hand, Proctor argues that technology, far from the solution, is part of the problem. She points out how information delivery technology reduces the quality of the social experience. This is quite significant in that, as Katz, Haley Goldman and Foutz illustrate, the visit to the museum is often more of a cultural ritual and opportunity for a shared group experience than it is for pure information transfer from the "resource" (museum) to the "customer" (visitor). Like Proctor, Ling and Murero warn about the potential distraction that handheld devices bring into the museum environment. They do so by highlighting studies that show that museum visits are considered a social activity and that handheld guides separate visitors by putting them into their own cognitive space, leaving little or no room for social interaction. At the very least, they create visual and perceptual barriers to easy interaction. Proctor criticizes the fact that the guides may block visitors from experiencing the exhibition halls and galleries in new and inventive ways as all the visitors follow the same preselected tracks and tours. In this respect, however, Timpson and Hulser and Bull point out that the most recent personal technologies have the ability to create individualized solutions. The ubiquitous nature of the internet, accessible through mobile technologies, now allows visitors to influence their exhibit route and even alter or create content for future participants in ways that were not possible even a few years ago.

What works?

It is unfortunate, but there is no simple solution to engaging previously unengaged audiences with museum content. Much hope was placed in earlier generations of communication technology, such as the radio, TV and computer to achieve this. No less than the mobile phone and the internet, each of these have also fallen short of the high hopes that were invested in them. However, the fact that most communication technology initiatives fall well short of the mark does not mean that there are not some valuable successes. In the studies included in this volume, we can report the success of *Ego Trap*, and several projects of Hulser and Bull. Ladly alone describes over eleven locative art exhibits that span the globe from as early as 1991. These projects cross both national and cultural lines (i.e., both art and science institutions) as well as reach beyond the walls of the institution. Threads that seem to tie these success stories together are that they successfully appeal to the individual, but allow the individual a large degree of interactive control. Power to the people! Also, for highly motivated individuals, mobile devices seem to work well when it comes to providing additional background to exhibits, such as is the case with art museums. Science museum visitors often come with the expectation that the exhibit should "do" something, and thus the tranquil interaction with a mobile phone is not that absorbing of attention. So this would suggest a future of gripping interaction with user choice.

Other opportunities created through the implementation of new technologies are the ability to connect with patrons at all stages of the museum experience. This includes before the site visit (which includes selecting the museum in the first place or planning the trip to the facility), as well as during the time on-site and post museum follow-up. As Zimmerman points out, these options are consumed by the enthusiasts; however, how can the use of mobile technologies also serve to retain and improve the experience of those with less patience or prior interest? By having individualized interaction, those with short attention spans can potentially be successfully engaged. This fact connects with the idea of *edu-tainment* of Kahr-Højland and Hug. Interactivity also allows those with a deeper interest to move beyond the exhibited material, moving them to access information and visual resources that are not on display. It further is designed to intellectually and affectively engage them.

Because of mediated communication technology, one can store a great deal of information and make it available, especially to those offsite. One can enrich and tag material as well as make it seem to appear in various revelatory contexts. These might be life-size, proportionate, or even layered or enhanced. Such virtual enhancements might include a dinosaur skeleton having virtual organs and flesh superimposed upon it, or hypothesized vocalizations. Hulser and Bull point out that at times, the medium in which these images are presented to visitors (for instance, mobile phone screens moved over an object) can provide a more accurate or meaningful representation than could be done in the context of the original artifact, which would be circumscribed by physical limitations.

Regarding future directions, several chapters focus on the need to better understand how mobile devices and social media can extend the learning process through time and space. The technologies can advance visitor understanding by allowing visitors to tie what they learn within the physical constraints of the institution to knowledge and experiences acquired outside of this setting. They also allow all kinds of people to engage the institution without visiting it at all. This potential points to a complete rethinking of pedagogical methods, with attention given to new approaches that integrate learning across multiple online and physical environments.

Scaffolding

Mediated communication technology may have its greatest impact in terms of scaffolding, that is, building a base of interest and knowledge in the visitor upon which more sophisticated understandings can be developed. This argument was advanced by several authors including Ling, Sanchez, and Hulser and Bull. Caron and Caronia also describe the importance of the semantic and pragmatic scaffolding that was provided by the co-text surrounding exhibits. They point out how this supposedly frames an interpretive path and establishes the boundaries of the visitors' sense making, but that the process is at best formative and incomplete. Similarly Kahr-Højland illustrates the need for directional guidance, a narrative pathway, to make the goals of learning in a museum substantially greater than those of a visit to an amusement park. This theme was also echoed by Ling. Moreover, Zimmerman highlights how we no longer need to think in terms of building physical structures or even singular institutions (or technologies contained within those institutions) to foster a continuity of learning. This is because we now have technology that can scaffold that connection for us.

Micro-steps in comprehensive, reciprocal learning processes

Many contributors acknowledge that those wishing to use mediated communication technologies effectively must overcome several hurdles, including cost, intellectual property rights, organizational conflict, and time investment.

Intellectual property remains a hurdle for easy distribution of materials. Security comes into play as museum websites and applications are a prime target for those seeking a site for attack or malicious re-use. Also important is the possible misuse by ordinary users. Because of liability issues, among others, unauthorized use or even digital material acquisition is an important consideration. Vigilant patrolling by site owners of user-submitted materials seems an essential requirement for any virtual areas that allow public accessibility. This is true especially in terms of photographs or videos. The privacy of users can also be compromised by their use. Children and teens can offer too much individual information that can be used against them. For any public institution, there are minimal (if not higher) standards of taste and propriety. Problems of staying on topic, though not as pernicious perhaps as the foregoing issues, remains another area that requires attention and monitoring if the public-input materials are to remain generally useful.

Control is also important in that topics can be sensitive to certain groups, as is the case of the Canadian Human Rights Museum. The role of the curatorial staff is important in making judgment in this regard. This becomes particularly confounded when users are able to post material that can be construed (inaccurately) as being "sourced" by the museum. Another concern that remains relevant is the potential for so-called Disneyfication of museums altogether. Caron and Caronia, as well as Kahr-Højland, discuss the disadvantages when opening production of content to all consumers. As discussed in an earlier section, such tendencies raise serious questions about the role and responsibility of a museum to act as a filter as well as the expertise of a curator. Ling ponders whether everything should be reduced to the logic of a theme park: the rubric would appear to be that as long as material is entertaining, it can stay. If this is the standard, one can join Zimmerman in inquiring about the fate of one of the original impulses of the movement to establish museums, namely to serve to enlighten the public. Most museums *qua* institutions were built to serve the educational, cultural, and aesthetic needs of the public, not to become houses of entertainment. In this context, mediated communication technology can appear to be as problematical as much as a panacea.

Current state of research and technology

As described by various authors in this collection, increasing numbers of museums of various types have been attempting to use digital technologies to enhance the visitor experience. However, it is imperative that we realize the limits of mobiles in exhibits. As Sanchez comments, with all efforts being put into attaining new, younger markets, institutions should not forget that no product can provide everything to everyone. In fact, often these endeavors run the risk of alienating the current target audience. Sanchez contrasts museums with Second Life in order to demonstrate commitment to the group and how only through shared values and desires can the group be cohesive and interested in the same experiences. However, like Second Life, museums can (and are) redefining themselves, allowing for greater customization. In her chapter, Murero also addresses concerns for, and proposes a method for, evaluating the risk and opportunities for innovations in the areas of social media and mobile communication technology.

The current state of theory in the museum environment involves a vision of how – and how not – to think about mobile devices and social media. Many of these highly debated points have been the focus of research in the mobile technology field. Recently, two such studies were conducted at Rutgers University through the Center for Mobile Communication Studies, the world's first academic unit to focus solely on social aspects of mobile communication. Because of the way in which they bolster the key findings of the chapters in this volume, these two studies will be described briefly, and then their findings will be summarized.

The first study by Youn and Titus, titled *Mobile Phones in U.S. Science Museums* (2009), surveyed museums in the United States concerning their use of mobile phones as interactive vehicles in their exhibits. Those science museums that were using mobile phone programs were analyzed in terms of the motivation for using this technology, how the programs were working, and the impact of these programs. The second study, *Looking Towards the Future of Mobile Learning through the Eyes of Experts* by Bridges, Jung, Lee and Shic (2009), explored the present status and anticipated future of informal education via mobile learning technologies in science museums. Data was collected via interviews with museum and related industry experts. Researchers examined:

- the reception to date in terms of incorporating mobile devices as a feature of interactive exhibits;
- anticipated steps in this area;
- and how the experts foresaw the future of mobile

> learning and informal science learning, especially in terms of enhancing visitors' experience in museums.

Although the detailed results cannot be presented here, these two interrelated studies found that mobile phone technology in U.S. science museums is still very much alive and well. In terms of the museum survey, it is clear that mobile technology continues to be an emerging trend, and many museums plan to exploit mobile technology in the days ahead. Yet the museums told the researchers that while they continue to believe the technology will have an ever-growing impact, at this point it has not delivered on its promise in terms of either connecting to audiences or moving content to users. Although only a minority of the sampled museums has currently integrated mobile devices and tools into their exhibits, the museums reported that there continues to be great interest in the implementation of digital technologies by museums that have not yet adopted the technology. Indeed, among non-using museums, over 60% of those surveyed have plans to introduce the technology at some near time (Youn and Titus, 2009).

In terms of the survey, the researchers also found that the experience of mobile communication technology for informal learning has not lived up to the optimistic predictions. They also see expanding adoption of the mobile platform and believe firmly that the mobile's impact is more positive than negative. Much like LaBar, Timpson, and Zimmerman expressed, the surveyed experts seemed to agree on the necessity of incorporating mobile learning into the museum setting (Bridges, Jung, Lee and Shie, 2009).

In all, these surveys reinforce the idea that mobile communication is making a difference, albeit less than anticipated. The early lack of definitive success has not dimmed the optimism for future potential as the technology matures and the audience cohorts change and the installed base of technically adept users grows. These findings, combined with the chapters discussing mobile communication and social media, suggest that the museum community's appetite is undiminished by the absence of meteoric success. Though problems remain, given the obvious potential of the technology, and some of the promising cases discussed in this volume's chapters, the assessment is eminently reasonable.

Conclusion

As illustrated throughout the chapters in this volume, the increasing efforts made by museums and science centers to engage existing and new audiences through mobile media and social networking tools have been met in some cases enthusiastically, but in many cases less so. Institutions have had various aims, from making their holdings and exhibits ever-more available to audiences, to increasing engagement and response by broad audiences to exhibit materials, perhaps looking for ways to build a bond between the museum qua an institution and the visitor. Again, there have been some successes, but also many less-than-stellar outcomes. In some cases, museums have sought not only interaction but also co-creation of materials by audience members. This in turn breaks down the walls between creators of knowledge and recipients, which can both enrich and create chaos for museum experience and the social role these institutions uphold. Although not dramatic in terms of numbers, such initiatives have yielded some intriguing results.

Beyond the mechanisms of specific programs described in these chapters have been the larger aims of museums, and here the aim has been much deeper than operational projects and has instead been aimed at the social role of the museum as an agent of social change. At this level, museums have sought to make their collections, and the insights of staff, more widely available to the public, especially those sectors that have traditionally ignored or by-passed such cultural institutions. These initiatives unfortunately come at a time when many museums struggle with pressing financial demands. They must weigh the return on investment of such costly programs, which is difficult to measure in the short term. Such assessments often have difficulty disentangling the attraction and currency of social media and mobile communication with other motives and potential outcomes. Beyond the accountant's spreadsheet, it is clear that while digital technologies have done much to connect museums with visitors, and allow institutions to have a stronger grasp on their internal resources, there continues to be an undertone of fear that more up-to-the-minute social media and digital tools will trump current technologies.

Thus overcoming the fear factor becomes a critical factor in deciding to make a leap into the unknown. Just as is the case with newspapers and the internet, museums may not know if they will benefit and prosper from going the route of social media and mobiles, yet they certainly know that if they do not a dismal fate awaits. If a museum is not investing in technology to make resources available, functionally speaking it is in danger of being by-passed or even becoming extinct. In terms of web presence, this is certainly the case. It is difficult to conceive that a museum would not have a website, for instance, even it were quite prosaic.

Yet for all the excitement and promise of social media and mobile communication, many in the traditional museum community have a strong anti-technology feeling. Not far below the surface is the often voiced (albeit *sotto voce*) view that the digital world generally, and that of social media specifically, oppose the cultural role of museums. Indeed when one considers the content of many Twitter or Instant Message (IM) comments, it is understandable why some would hunger for the profound essay. For such people, the correlate is that the museum should not be too much a part of contemporary society but should always remain loyal to the grand mission of intellectual progress. Though one eye should be kept on serving contemporary society, the other should be kept firmly on the enduring and ever-enriching task of fostering high culture and scientific knowledge, an enterprise that stretches from ancient times into the distant future.

Assuming this analysis is correct, what is the likely road

ahead? There is no putting the genie of social media/mobile communication back in the bottle. The idea is simply quaint and for the foreseeable future an operational nonstarter. (The one exception to this assertion is that some attempts might be made to mount exhibits without social media/mobiles, but done in an attempt to show what "has been lost" or "what life used to be like." Thus these would by analogy be what living farm museums are to today's agribusiness.) And like other technologies, the consequences of social media/mobile communication have both positive and negative components. Nevertheless, for museums and science centers it seems that the outcomes are largely positive in that they are able to attract new audiences and strengthen their hold on old ones. As well, the bar of ambition is set too high. Some of the very characteristics that make social media/mobiles attractive to use work against their applicability to museums, which often emphasize the rewards of accepting and surmounting difficult challenges rather than the immediate gratification of impulse. Yet if properly deployed, as demonstrated by *Ego Trap*, these technologies can stimulate and connect young people to the museum, making easier future *entrée* to the fascinating riches offered by museums. As suggested by several chapter authors, by combining the interest in fun and social competition, the social media/mobile communication nexus can offer an important springboard to achieving the complex and multi-layered goals of the museum and science center as vital cultural institutions. It may not be the best of all possible worlds, but it is one that offers excitement and opportunities aplenty. And social media and mobile communication technologies do seem to already have captured the full attention of several billion people.

Index

About the Authors

Lora Appel

Lora Appel is a third semester student in the MCIS program at Rutgers University, and a research assistant to Dr. James Katz, the chair of the department of Communication and Information. She holds a fellowship with Ethicon, Johnson & Johnson. Lora graduated with an International Bachelor of Business Administration from the Schulich School of Business in Toronto, Canada, and furthered her studies earning a postgraduate certificate in Corporate Communication at Seneca College. She studied on exchange at the ESADE School of Business in Barcelona and at Ghent University in Belgium. Before starting her studies at Rutgers, Lora completed an internship with the Canadian Institute of Chartered Accountants and helped organize the International Pediatric Rehabilitation conference in association with the Universitair Ziekenhuis Ghent Hospital.

Steve Bull

After graduating from NYU's Interactive Telecommunications Program, Steve Bull founded Cutlass, an award winning locative media studio. Since 2001, Cutlass' mobile tour guides have been available in Canada, the UK, and the US, including niche versions Quirky World and Hollywood-USA. The *New York Times* lauded the launch of the first tour in Greenport, NY. The New-York Historical Society commissioned *Slavery In New York*, a series of mobile tours that were sponsored by the Verizon Foundation and the US Department of Education in order to extend the museum's physical reach and attract new audiences. Hollywood-USA received a finalist award for TeleAtlas: Maps in Apps in 2007 and was a finalist at the 2009 NAVTEQ Global LBS Challenge. With Hana Iverson, Steve designed *Neighborhood Narratives* for Temple University, and the mobile phone experience for Iverson's Cross/Walks. He also worked in development at Interval Research and designed user action and the prototype for the Experience Museum in Seattle.

André H Caron

André Caron is Full Professor in the Communication Department (Canada Université de Montréal), Director of the Bell Chair in Interdisciplinary Research on Emerging Technologies and serves as founding Director of the GRJM/CYMS (Centre for Youth and Media Studies). His fields of specialty include: diffusion of new technologies and their impact on various social environments; formative and summative research on youth and the media and international cultural and policy media issues. His research interests include: the appropriation of mobile culture in society and the analysis of interactive media systems.

Letizia Caronia

Letizia Caronia is Professor at the Faculty of Education (University of Bologna, Italy) where she teaches Philosophy of Education and Epistemology and Methods in Qualitative Research. She is visiting scholar at the Centre of Interdisciplinary Research On Emerging

Technologies (Department of Communication, University of Montreal) where she currently leads research projects on the appropriation of information and communication technologies in everyday life. Her research deals with the role of everyday language, interaction and culture in children socialization in the family and at school. Letizia has authored and co-authored more then fifty chapters in collective books and articles in national and international scientific journals in Italian, English and French.

Susan Foutz

Susan Foutz is a Senior Research Associate at the USA Institute for Learning Innovation. Since joining the Institute in 2003, Susan has co-edited the books *In Principle, In Practice* and *Free-Choice Learning and the Environment* and coordinated the related conferences. Her research interests include Positive Youth Development, family learning in museum exhibitions and programs, and the use of technology and media in the pursuit of free-choice learning. She holds a MA in Museum Studies from the University of Nebraska and a BA in Sociology/Anthropology from Ohio Wesleyan University.

Kate Haley Goldman

Kate Haley Goldman recently joined the senior staff of the Center for Interactive Learning, a non-profit founded by the Space Science Institute. Prior to taking this position as Director of Learning Research and Evaluation, she was a Senior Research Associate at the Institute for Learning

Innovation since 2000. Her work concentrates on furthering theory and practice of the use of technology in museums and related informal learning environments. She has directed projects both in the US and abroad, involving exhibits and program evaluation, mobile phones and smartphone apps, websites, gaming, augmented and mixed reality, novel data visualization systems, and online learning. Recent projects include audience research for the *Encyclopedia of Life*, summative evaluation of the *WolfQuest* game, program-level evaluation of NOAA's *Science on a Sphere*, and the newly-awarded National Science Foundation projects *STAR_Net* and *Making Space Social*. She is Co-PI of the NSF-funded open source project Open Exhibits. (OpenExhibits.org) She is particularly interested in developing reliable and valid evaluation methodologies for online environments and technologies and deepening the theoretical knowledge base as it applies to technology and these environments. She has a bachelor's degree in Anthropology from Bryn Mawr College and has extensive training in educational measurement, statistics and evaluation.

Nanna Holdgaard

Nanna Holdgaard is a PhD Student at the IT University of Copenhagen. For her PhD project she is examining Danish online museum communication from a user perspective with particular focus on the use of social media. She has both theoretical and practical experience with digital museum communication, holds an MSc in Digital Design and

Communication, and has worked at several Danish museums.

Theo Hug

Theo Hugg is professor of educational sciences at the University of Innsbruck and coordinator of the Innsbruck Media Studies research forum. His areas of interest include media education and media literacy, e-education, m-learning and microlearning, theory of knowledge and philosophy of science. He is the author and/or editor of several books on various aspects of media, communication and education.

Kathleen Hulser

Kathleen Hulser has been public historian at the New-York Historical Society since 1999. She is currently working on two upcoming exhibition projects. *Nueva York*, a look at the influence of the Spanish speaking world on New York for three centuries, opens at El Museo del Barrio in 2010. *Revolutions: America, France, Haiti* will be the exhibition celebrating the grand re-opening of the New-York Historical Society in 2011. Kathleen also teaches history at the New School and at New York University's Department of Social and Cultural Analysis. Her public programming projects include a walking tour of *Noted and Notorious Women of Lower Manhattan* and a cellphone/ iPod tour of *Hidden Sites of Slavery and Freedom* (www.nyhistory.org/podcasts). She also recently produced a media short on the Marquis de Lafayette, the French Revolution, and the Guillotine.

Anne Kahr-Højland

Anne Kahr-Højland is curator and process leader at the Viking Ship Museum in Roskilde, Denmark, where she is head of a process of change as regards the concept of facilitation and interpretation at the museum. Anne has a background as a researcher in digital facilitation of learning within museums where she especially has been working with Design Based Research. As part of her doctoral studies at DREAM/ University of Southern Denmark, Anne was the principal organizer of a cellphone-facilitated narrative at the science center Experimentarium. This exhibition, called *EGO-TRAP – the cellphone is your key*" opened in a test version in spring 2007. It was designed as a learning resource for pupils in Danish upper secondary schools as a digital narrative – a kind of computer-assisted role-playing game where the visitor takes a guided tour of the exhibition through his / her own mobile phone.

James E Katz

James E Katz is professor and chair of the Department of Communication at Rutgers University where he also directs the Center for Mobile Communication Studies. He holds the rank of Professor II, Rutgers' highest professorial rank, which is reserved for those who have achieved national and international eminence in their field. Professor Katz has devoted much of his career to exploring the social consequences of new communication technology, especially the mobile phone and internet. Currently, he is looking at how personal communication technologies can be used by teens from urban environments to engage in informal science and health learning.

Emlyn H. Koster

Initially a geologist engaged in research and teaching, Emlyn Koster's focus shifted to the greater possibility of museums in the natural history and science-technology realms to illuminate the challenges and opportunities surrounding society and the environment. After launching the Royal Tyrrell Museum of Palaeontology in Alberta, and overseeing a content and profile facelift of the Ontario Science Centre in Toronto, he became President and CEO of Liberty Science Center in 1996. Here, he has been at the helm of its journey of increasing regional relevancy, major expansion and total renewal, 2007 reopening, and in the assembly of steering colleagues to plan and implement this unusual scope and pace of transformation. His service to the field includes chairing the board of the Maryland-based Institute for Learning Innovation. His honors include a Chevalier Medal from France's l'Ordre des Palmes Academique, Humanitarian Award from the American Conference on Diversity, and Chairman's Award from the New Jersey Research and Development Council for the state-wide educational impact of Liberty Science Center. Amplifying his several articles annually that shape the advancing philosophy and practice of relevancy in the science museum field, invited keynotes have been given across Canada and the United States, as well as in Australia, China, Colombia, Costa Rica, Japan, Mexico, and Thailand. His media profile includes The *New York Times* and The *Wall Street Journal*.

Wayne LaBar

As Vice President of Exhibitions and Featured Experiences, Wayne LaBar oversees thematic content, exhibitions, large format film and 2D and 3D overall design at Liberty Science Center (LSC). As Principal-in-Charge of LSC Experience Services, which he launched and leads, LSC helps design and develop science centers and museums around the world, as well as produces travelling exhibitions that tour internationally. LSC's experiences are recognized as having succeeded in making the Center's informal experiences more relevant to guests' everyday lives and global issues; educating and inspiring action; addressing ongoing developments in science and technology; and enhancing the Center's learning and aesthetic environment – redefining the role of a modern science center. As part of both positions, Wayne explores how new technology can encourage guests to interact with exhibitions and the exhibition process. Examples are *Science Now Science Everywhere* and *SNSE*, a unique cellphone application; *Exhibit Commons*, an initiative that invites guests to contribute to the Center's mutable exhibitions; and implementation of social media software that enables the public to get involved in the development of an exhibition on the science of cooking (http://cookingexhibitchefs.ning.com/).

Martha Ladly

Martha Ladly is an associate professor and researcher at OCADU in Toronto, Canada, where she is director of the Interdisciplinary Master's Program in Art, Media and Design, and the Chair of the Research Ethics Board. She specializes in technology in art and design, and is a senior researcher with the OCADU Mobile Experience Lab. She was formerly the Director of HorizonZero (www.horizonzero.ca), the Canadian digital arts and culture website. Martha also had a prominent international recording career as a musician, performer, and writer. She earned two certified gold discs, numerous top-ten singles and albums, and a Juno award. Throughout the 1990s Martha worked with Peter Gabriel and Real World, leading design teams for award-winning interactive arts and entertainment projects. She is an author and the co-editor of the anthology *Mobile Nation: Creating Methodologies for Mobile Platforms.*

Rich Ling

Rich Ling is a professor at the IT University of Copenhagen, Denmark and has a position as a sociologist at the Telenor research institute located near Oslo, Norway. He has been the Pohs visiting professor of communication studies at the University of Michigan in Ann Arbor, Michigan and now he holds an adjunct position in that department. He is the author of the soon to be published book: *New Tech, New Ties: How mobile communication is reshaping social cohesion* (MIT). He is also the author of a book on the social consequences of mobile telephony entitled *The Mobile Connection: The cell phone's impact on society* (Morgan Kaufmann) and along with Per E. Pederson the editor of the book *Mobile Communications: Renegotiation of the Social Sphere* (Springer). Along with Scott Campbell he is the editor of *The Mobile Communication Research Series* and he is an associate editor for *The Information Society*, as well as *Norsk Medietidsskrift*.

Ellen Lynch

Ellen Lynch is the Director of Exhibition Development and Operations at Liberty Science Center. She has been a member of the exhibitions team there for over 19 years and has extensive experience in managing, developing and implementing interactive exhibition experiences. Currently Ellen oversees the content creation team as well as exhibit operations, which are heavily based in technology applications. Recent projects include co-management and implementation of the Center's *Science Now, Science Everywhere* mobile learning initiative.

Monica Murero

Monica Murero is the Director of the E-Life International Institute, founded in 2003 at the University of Firenze – Center of Excellence in Media Integration and Communication (MICC). She is Associate Professor in Sociology of New Technology, and in Politics of e-Government at the University Federico II. An international award-winner for scientific excellence, Monica is the beneficiary of the Rientro Cervelli Award (*Brain Gain*), the most prestigious Italian prize for distinguished international academics (MIUR, the Italian Ministry of University and Research). She is an Honorary Lifetime Member of AoIR, the International Association of Internet Researchers (www.aoir.org), Member of the AoIR Executive Committee (2002/2009), on the Board of Directors (Treasurer, 2005/2009), and Vice President Candidate in the 2009 AoIR elections. Her research interests revolve around the intersections among new technologies, new media, and psycho-social change. In particular, she is interested in the interdisciplinary field of internet research and media integration, the effects of innovation upon human behavior, and its steady encroachment into crucial aspects of everyday life (i.e., e-health, e-museums, e-learning, etc.). She is a consultant to the World Health Organization and the European Commission.

Nancy Proctor

Nancy Proctor heads up mobile strategy and initiatives for the Smithsonian Institution. With a PhD in American art history and a background in filmmaking, curation and art criticism, Nancy published her first online exhibition in 1995. She co-founded TheGalleryChannel.com in 1998 with Titus Bicknell to present virtual tours of innovative exhibitions alongside comprehensive global museum and gallery listings. TheGalleryChannel was later acquired by Antenna Audio, where Nancy headed up New Product Development from 2000-2008, introducing the company's multimedia, sign language, downloadable, podcast and cellphone tours.

She also led Antenna's sales in France from 2006-2007, and worked with the Travel Channel's product development team. From 2008-2010 she was Head of New Media at the Smithsonian American Art Museum. Nancy is program chair for the Museums Computer Network (MCN) conference and co-organizes the Tate Handheld conference and the regular DC Mobile Meet-ups for local area cultural professionals. She also manages MuseumMobile.info, its wiki and podcast series, and is Digital Editor of *Curator: The Museum Journal*.

Corey Timpson

Corey Timpson is responsible for managing design and new media initiatives related to, and supporting all aspects of, the Canadian Museum for Human Rights (CMHR)'s collections, research, exhibits, and corporate. His primary focus at the CMHR is to facilitate interactions and dialogue between and among visitors (on-line and in-house) through the use of new media and digital technology relying on sustainable, scalable, and efficient data and interaction models. Prior to accepting his role at the Canadian Museum for Human Rights, Corey spent eight years at the Canadian Heritage Information Network and the Virtual Museum of Canada where he led the interface and creative design, information architecture, and web management teams. Outside of the museum office, Corey can be found working on personal digital art, photography, and digital design projects or consulting on creative design and HCI projects.

Bjarki Valtysson

Bjarki Valtysson is an Assistant Professor in the DCMC group (Design – Culture – Mobility – Communication) at the IT University of Copenhagen. Research interests include cultural, media and communication policies, social media, remix culture and, in particular, the intersections between digital communication and cultural institutions.

Tim Zimmerman

Tim Zimmerman is an Assistant Professor of Science Education at Rutgers University. His research seeks to improve our understanding of student learning of ocean and environmental science concepts, broadly speaking, as a means for improving science education and scientific literacy. In particular, he focuses on the design and use of internet technologies (e.g., real-time data, GIS, wiki's, social-networking, internet media, etc.) in classrooms and on field trips (museums, aquariums, outdoor settings, etc.) for both traditional computers and on mobile computing devices. These tools are then used to research the nature of learning across formal and informal contexts; the connections between science learning and environmental decision-making; and the learning and teaching of ocean science as a progression of conceptual understanding.

Also from MuseumsEtc

Science Exhibitions: Curation and Design

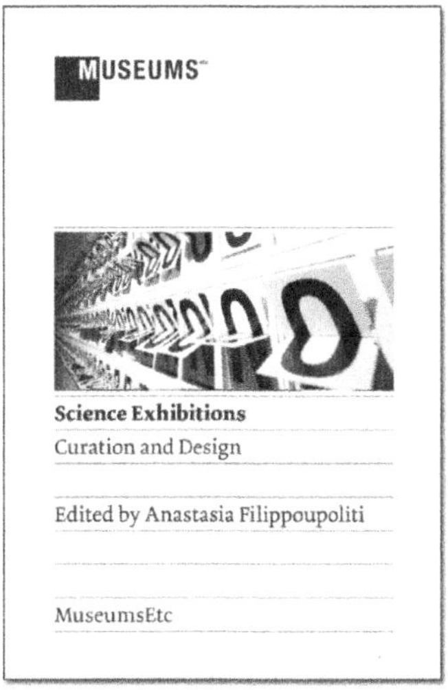

ISBN: 978-0-9561943-5-0 [paperback]
ISBN: 978-1-907697-03-6 [hardback]
Order online from www.museumsetc.com

Edited by Dr Anastasia Filippoupoliti, Lecturer in Museum Education, Democritus University of Thrace, Greece

This important new book is the companion volume to Science Exhibitions: Communication and Evaluation and examines how best to disseminate science to the public through a variety of new and traditional media. With some 20 essays from leading practitioners in the field, and over 500 pages, it provides an authoritative, stimulating overview of new, innovative and successful initiatives.

The essays draw on cutting-edge experience throughout the world, and include contributions from Australia, Brazil, France, Germany, Singapore and New Zealand – as well as the UK and USA.

Rethinking Learning: Museums and Young People

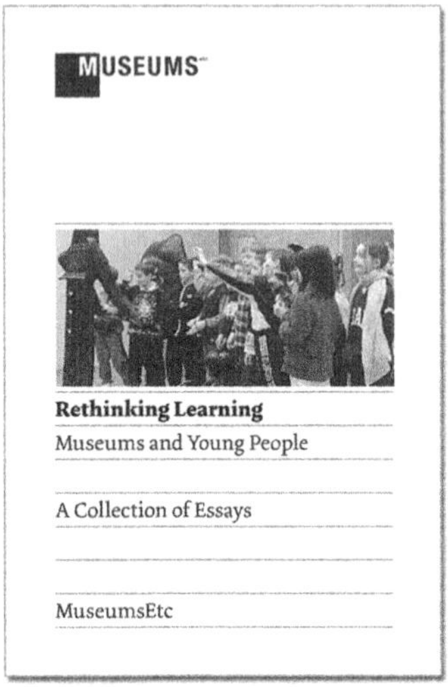

ISBN: 978-0-9561943-0-5 [paperback]
Order online from www.museumsetc.com

Practical, inspirational case studies from senior museum and gallery professionals from Europe and the USA clearly demonstrate the way in which imaginative, responsive services for children and young people can have a transformational effect on the museum and its visitor profile as a whole.

Authors recount how, for example – as a direct result of their focus on young people – attendance has increased by 60% in three years; membership has reached record levels; and repeat visits have grown from 30% to 50%. Many of the environments in which these services operate are particularly challenging: city areas where 160 different languages are spoken; a remote location whose typical visitor has to travel 80 miles.

This is an essential book not only for those working with children and young people, but for those in any way concerned with museum and gallery policy, strategy, marketing and growth.

The Power of the Object: Museums and World War II

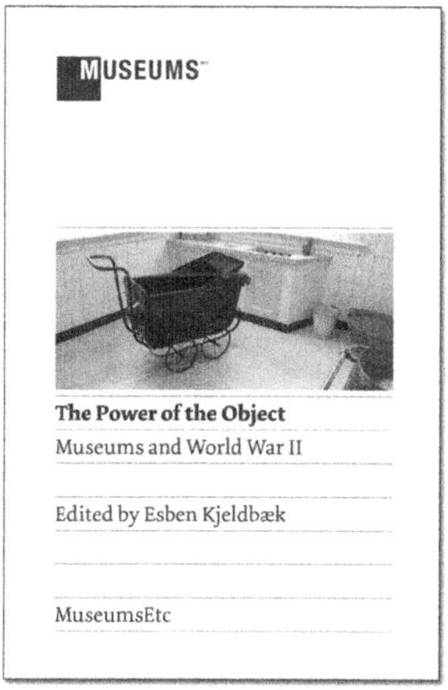

ISBN: 978-0-9561943-4-3 [paperback]
Order online from www.museumsetc.com

Edited by Esben Kjeldbæk, Head of the Museum of the Danish Resistance 1940-1945, National Museum of Denmark

In this important new book, based on a conference held by the National Museum of Denmark, international museum professionals deal with the key issues affecting all history museums, using as the basis for their insights the interpretation by museums of World War II.

Among the many issues the contributors address are:
How best can abstractions like cause, effect and other ideas be interpreted through objects?
Just how is the role of objects within museums changing?
How should we respond when increasingly visitors no longer accept the curator's choice of objects and their interpretation?
How can museums deal effectively with controversial historical issues?

These essays explore how history museums can help explain and interpret the thinking of past generations, as well as their material culture.

Narratives of Community: Museums and Ethnicity

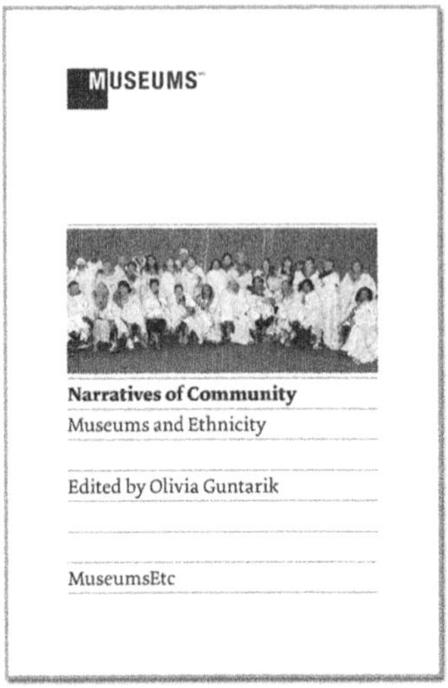

ISBN: 978-1-907697-05-0 [paperback]
ISBN: 978-1-907697-06-7 [hardback]
Order online from www.museumsetc.com

Edited by Dr Olivia Guntarik, RMIT University, Melbourne, Australia

In this groundbreaking book, cultural theorist and historian Olivia Guntarik brings together a collection of essays on the revolutionary roles museums across the world perform to represent communities. She highlights a fundamental shift taking place in 21st century museums: how they confront existing assumptions about people, and the pioneering ways they work with communities to narrate oral histories, tell ancestral stories and keep memories from the past alive.

The philosophical thread woven through each essay expresses a rejection of popular claims that minority people are necessarily silent, neglected and ignorant of the processes of representation. This book showcases new ways of thinking about contemporary museums as spaces of dialogue, collaboration and storytelling. It acknowledges the radical efforts many museums and communities make to actively engage with and overthrow existing misconceptions on the important subject of race and ethnicity.

New Thinking: Rules for the (R)evolution of Museums

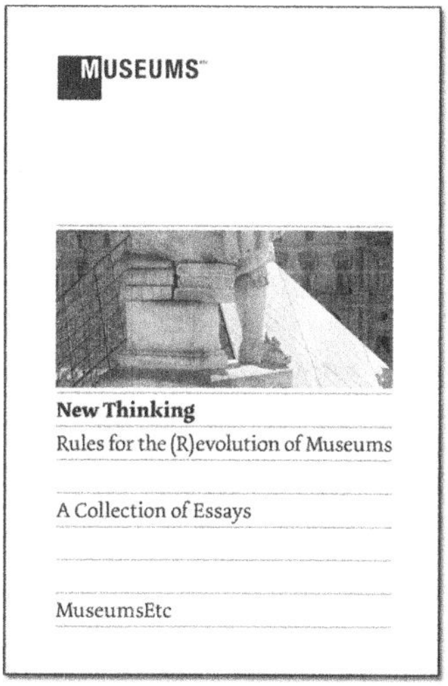

ISBN: 978-0-9561943-9-8 [paperback]
ISBN: 978-1-907697-04-3 [hardback]
Order online from www.museumsetc.com

This new collection of essays by leading international museum practitioners focuses on the across-the-board innovations taking place in some of the world's most forward-thinking museums – and charts the new directions museums will need to take in today's increasingly challenging and competitive environment.

Among the twenty world-class organisations sharing their innovative experiences are:

- Canada Agriculture Museum
- Canadian Museum for Human Rights
- Conner Prairie Interactive History Park
- Cooper-Hewitt National Design Museum
- Imperial War Museum
- Liberty Science Center
- Miami Science Museum
- Museum of London
- National Museum of Denmark
- Royal Collection Enterprises
- Smithsonian American Art Museum
- Victoria & Albert Museum
- Wellcome Collection

The New Museum Community

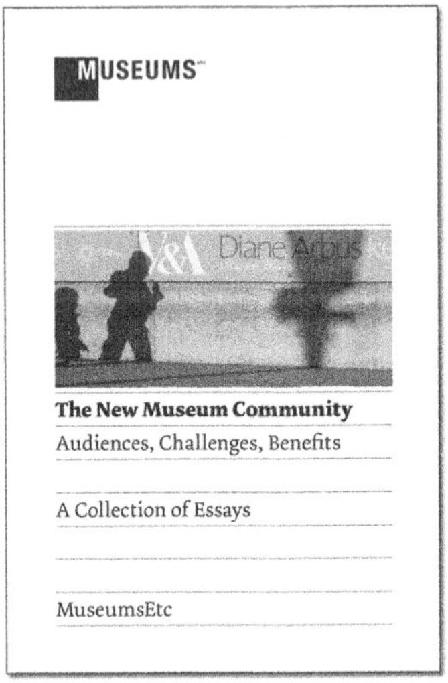

ISBN: 978-0-9561943-7-4 [paperback]
ISBN: 978-1-907697-02-9 [hardback]
Order online from www.museumsetc.com

In a series of authoritative essays, this important new book examines in depth how museums globally are succeeding – in many different ways – in reaching and involving social groups traditionally overlooked or excluded from the museum experience. The New Museum Community is an invaluable and inspiring guide for any museum, gallery or cultural organisation intent on engaging with the broadest possible audience.

Among the world-class organisations sharing their practical experience are:

- British Museum
- Manchester Museum
- Miami Science Centre
- Museum of London
- Museum Victoria
- Museo Nazionale della Scienza, Milan
- Penn State University
- Tate Britain
- University of Brighton
- University of Edinburgh
- University of Newcastle

Creating Bonds: Marketing Museums

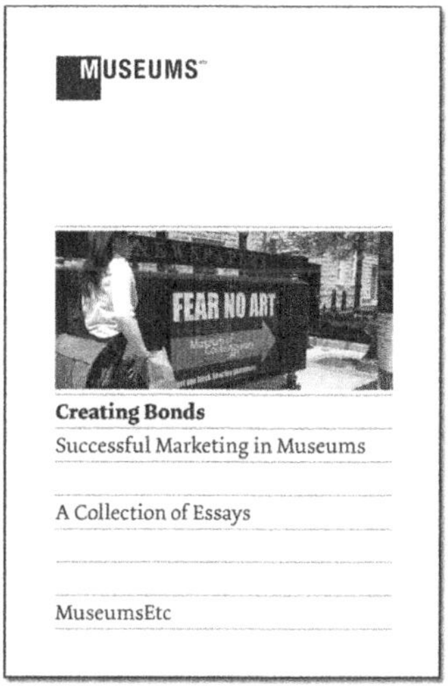

ISBN: 978-0-9561943-1-2 [paperback]
Order online from www.museumsetc.com

Marketing professionals working in and with museums and galleries throughout the UK, USA and beyond, share their latest insights and experiences of involving and speaking to a wide range of constituencies. Their practical, insightful essays span the development and management of the museum brand; successful marketing initiatives within both the small museum and the large national museum environment; and the ways in which specific market sectors – both young and old – can be effectively targeted.

Contributors include: Amy Nelson, University of Kentucky Art Museum; Aundrea Hollington, Historic Scotland; Adam Lumb, Museums Sheffield; Bruno Bahunek, Museum of Contemporary Art, Zagreb; Claire Ingham, London Transport Museum; Kate Knowles, Dulwich Picture Gallery; Danielle Chidlow, National Gallery; Margot Wallace, Columbia College; Rachel Collins, Wellcome Collection; Sian Walters, National Museum Wales; Gina Koutsika, Tate; Alex Gates, North Berrien Historical Museum.

Colophon

Published by
MuseumsEtc Ltd
Hudson House
8 Albany Street
Edinburgh EH1 3QB
United Kingdom

www.museumsetc.com

The paper used in printing this book comes from responsibly managed forests and meets the requirements of the Forest Stewardship Council™ (FSC®) and the Sustainable Forestry Initiative® (SFI®). It is is acid free and lignin free and meets all ANSI standards for archival quality paper.

This book is typeset in Underware Dolly and Adobe Myriad Pro.

ISBN: 978-1-907697-11-1

A CIP catalogue record for this book is available from the British Library.

Lightning Source UK Ltd.
Milton Keynes UK
UKOW06f1808241115

263459UK00007B/327/P

9 781907 697111